APPLIED PROBABILITY

A Series of the Applied Probability Trust

Editors
J. Gani C.C. Heyde

Robert Azencott
Didier Dacunha-Castelle

Series of Irregular Observations

Forecasting and Model Building

Springer-Verlag
New York Berlin Heidelberg Tokyo

Robert Azencott
Université de Paris-Sud
Equipe de Recerche Associée
au C.N.R.S.
Statistique Appliquée Mathématique
91405 Orsay Cedex
France

Didier Dacunha-Castelle
Université de Paris-Sud
Equipe de Recerche Associée
au C.N.R.S. 532
Statistique Appliquée Mathématique
91405 Orsay Cedex
France

Series Editors

J. Gani
Statistics Program
Department of Mathematics
University of California
Santa Barbara, CA 93106
U.S.A.

C. C. Heyde
Department of Statistics
Institute of Advanced Studies
The Australian National University
Canberra, ACT 2601
Australia

AMS Classification 62-01, 62M10, 62M15

Library of Congress Cataloging-in-Publication Data
Azencott, Robert.
Series of irregular observations.
(Applied probability)
Bibliography: p.
Includes index.
1. Stochastic processes. I. Dacunha-Castelle, Didier. II. Title. III. Series.
QA274.A94 1986 519.2 86-1834

French Edition, *Series d'Observations Irregulieres*, © Masson, Editeur, Paris, 1984.

Printed and bound by R.R. Donnelley and Sons, Harrisonburg, Virginia.
Printed in the United States of America.

9 8 7 6 5 4 3 2 1

ISBN 0-387-96263-8 Springer-Verlag New York Berlin Heidelberg Tokyo
ISBN 3-540-96263-8 Springer-Verlag Berlin Heidelberg New York Tokyo

CONTENTS

INTRODUCTION

For the past thirty years, random stationary processes have played a central part in the mathematical modelization of numerous concrete phenomena.

Their domains of application include, among others, **signal theory** (signal transmission in the presence of noise, modelization of human speech, shape recognition, etc. ...), **the prediction of economic quantities** (prices, stock exchange fluctuations, etc.), **meterology** (analysis of sequential climatic data), **geology** (modelization of the dependence between the chemical composition of earth samples and their locations), **medicine** (analysis of electroencephalograms, electrocardiograms, etc. ...).

Three mathematical points of view currently define the use of stationary processes: **spectral analysis**, linked to Fourier transforms and widely popularized by N. Wiener; **Markov representations**, particularly efficient in automatic linear control of dynamic systems, as shown by Kalman-Bucy's pioneering work; **finite autoregressive and moving average schemes (ARMA processes)** an early technique more recently adapted for computer use and vulgarized by Box-Jenkins.

We have sought to present, in compact and rigorous fashion, the essentials of spectral analysis and ARMA modelization. We have deliberately restricted the scope of the book to one-dimensional processes, in order to keep the basic concepts as transparent as possible.

At the university level, in probability and statistics departments or electrical engineering departments, this book contains enough material for a graduate course, or even for an upper-level undergraduate course if the asymptotic studies are reduced to a minimum. The prerequisites for most of the chapters (1 - 12) are fairly limited: the elements of Hilbert space theory, and the basics of axiomatic probability theory including L^2-spaces, the notions of distributions, random variables and bounded measures.

The standards of precision, conciseness, and mathematical rigour which we have maintained in this text are in clearcut contrast with the majority of similar texts on the subject. The main advantage of this choice should be a considerable gain of time for the noninitiated reader, provided he or she has a taste for mathematical language.

On the other hand, being fully aware of the usefulness of ARMA models for applications, we present carefully and in full detail the essential algorithms for practical modelling and identification of ARMA processes. The experience gained from several graduate courses on these themes (Universities of Paris-Sud and of Paris-7) has shown that the mathematical material included here is sufficient to build reasonable computer programs of data analysis by ARMA modelling.

To facilitate the reading, we have inserted a bibliographical guide at the end of each chapter and, indicated by stars (*...*), a few intricate mathematical points which may be skipped over by nonspecialists.

On the mathematical level, this book has benefited from two seminars on time series, organized by the authors at the University of Paris-Sud and at the Ecole Normale Superieure (rue d'Ulm) Paris. We clarify several points on which many of the "classics" in this field remain evasive or erroneous: structure and nonstationarity of ARMA and ARMA seasonal processes, stationary and nonstationary solutions of the general ARMA equation, convergence of the celebrated Box-Jenkins backforecasting algorithms, asymptotic behaviour of ARMA estimators, etc.

We would like to thank the early readers of the French edition, particularly E. J. Hannan for his detailed and crucial comments, as well as M. Bouaziz and L. Elie for their thoughtful remarks.

Chapter I
DISCRETE TIME RANDOM PROCESSES

1. Random Variables and Probability Spaces

The experimental description of any random phenomenon involves a family of numbers X_t, $t \in T$. Since Kolmogorov, it has been mathematically convenient to summarize the impact of randomness through the stochastic choice of a point in an adequate set Ω (space of trials) and to consider the random variables X_t as well determined functions on Ω with values in $\mathbb{R}$.

The quantification of randomness then reduces to specifying the family $\mathcal{B}$ of subsets of Ω which represent relevant events and, for each **event** $A \in \mathcal{B}$, the **probability** $P(A) \in [0,1]$ of its occurrence.

Mathematically, $\mathcal{B}$ **is a σ-algebra**, i.e. a family left stable by complements, countable unions and intersections; P **is a positive measure** on $(\Omega,\mathcal{B})$ with $P(\Omega) = 1$, and the **real valued random variables** (r.v. for short) are the measurable functions $Y\colon \Omega \to \mathbb{R}$, that are such that $Y^{-1}(J) \in \mathcal{B}$ for any Borel subset J of $\mathbb{R}$

The main object here is the family X_t, $t \in T$, of r.v. accessible through experiment, and its statistical properties. The probability space $(\Omega,\mathcal{B},P)$ is generally built up with these statistical properties as a starting point, and in the present context plays only a formal part, useful for the rigor of the statements, but with no impact on actual computations.

In this text it will often be possible at first reading, to completely ignore σ-algebras and measurability. This should not hinder the understanding of the main results and their applications.

2. Random Vectors

For every numerical r.v. Y on $(\Omega,\mathcal{B},P)$, we write

$$E(Y) = \int_\Omega Y(\omega)dP(\omega)$$

for the **expectation** of Y, when this number is well defined.

If $X = (X_1 ... X_k)$, where the coordinates X_k are numerical r.v., we shall say that X is **a (real) random vector.** The **law of** X, also called the **joint law** of $X_1...X_k$ is the probability π on $\mathbb{R}^k$ defined by $\pi(A) = P(X \in A)$ for every Borel subset A of $\mathbb{R}^k$. One also says that π **is the image of** P **by** X. This is equivalent to

$$\int_{\mathbb{R}^k} f\, d\pi = \int_\Omega f \circ X\, dP = E[f(X)]$$

for every function f: $\mathbb{R}^k \to \mathbb{R}$ such that one of the three terms is well defined. Replacing $\mathbb{R}$ by $\mathbb{C}$, one defines simularly the **complex random vectors.**

The probability distribution of X is said to have a density φ: $\mathbb{R}^k \to \mathbb{R}^+$ (with respect to Lebesgue measure) when

$$E[f(X)] = \int_{\mathbb{R}^k} f\, d\pi = \int_{\mathbb{R}^k} f(x)\varphi(x)dx$$

for all f as above.

3. Random Processes

3.1. Definitions.

Let T be an arbitrary set. A **random process** X indexed by T is an arbitrary family X_t , $t \in T$, of random vectors defined over the same probability space $(\Omega,\mathcal{B},P)$, with values in the same space $E = \mathbb{R}^k$ or $\mathbb{C}^k$, called the state space of X. The set T often represents time; in particular when $T = \mathbb{N}$ or $T = \mathbb{Z}$,

X is called a **discrete time random process.**

Example 1. Such a model can be used to study the trajectory of a small particle in a liquid medium, moving under the action of random shocks due to molecular agitation. Here one chooses $E = \mathbb{R}^3$, and X_n is the position of the particle at the nth-observation.

Example 2. One has a finite sequence of numerical observations x_n, $1 \leqslant n \leqslant K$, for instance the daily temperatures observed at a precise geographic location over the course of two years. Assume there are enough irregularities in the graph of the sequence x_n, so that a stochastic model for this phenomenon seems like a good option. The point is to construct a process $(\Omega, \mathcal{B}, P, (X_n))$, with real valued X_n, such that it becomes "reasonable" (to be defined !) to say that the sequence x_n has been obtained by the random choice (according to the law P) of a point $\omega \in \Omega$, followed by the definition $x_n = X_n(\omega)$.

An important difference between Examples 1 and 2: it seems plausible to repeat experiment 1 in "identical" conditions; such an operation is much less plausible in type 2 cases. As we shall see, this is one of the reasons which suggest the use of stationary processes to study type 2 situations.

Paragraphs 3.2, 3.3, 3.4 may well be skipped at first reading.

3.2. Equivalent Processes*

Even in type 1 situations, where the random experiment can be repeated, the measurements can only bear on the positions $X_{t_1} \dots X_{t_N}$ effectively observed at fixed times $t_1 \dots t_N$. *At best*, the measurements can only supply a good approximation for the joint distribution $\pi_{t_1 \dots t_N}$ of $X_{t_1} \dots X_{t_N}$. Two stochastic models such that these joint distributions coincide for all $t_1 \dots t_N$ are hence **indistinguishable** in practice.

Two processes X, X' will then be called **equivalent** if the joint distributions

$$\pi_{t_1 \dots t_N} \quad \text{and} \quad \pi'_{t_1 \dots t_N} \quad \text{of} \quad X_{t_1 \dots t_N} \quad \text{and} \quad X'_{t_1} \dots X'_{t_N}$$

coincide for every finite family of indices $t_1 \dots t_N$. We now characterize the family of joint distributions of a process.

3.3. *Consistent Families of Distributions*

To each finite subset I of the set T of instants, associate the product space E^I, that is the set of finite sequences x_i, $i \in I$, where $x_i \in E$. For I,J finite subsets of T, with $J \subset I$, define Φ_{IJ}: $E^I \to E^J$ to be the natural projection, which amounts to forgetting the coordinates whose indices are not in J.

A family μ_I indexed by the finite subsets I of T, where for each I, μ_I is a probability on E^I, is said to be **consistent** if $\Phi_{IJ}(\mu_I) = \mu_J$ whenever $J \subset I$; this is equivalent to $\mu_I[\varphi_{IJ}^{-1}(A)] = \mu_J(A)$ for every Borel subset A of E^J.

3.4. *Theorem (Kolmogorov)*. *Let $E = \mathbb{R}^k$ or $\mathbb{C}^k$. Let T be an arbitrary set. For every finite subset I of T let μ_I be a probability on E^I. To guarantee the existence of a random process X indexed by T, with state space E, such that for all $I = \{t_1 \dots t_N\} \subset T$, μ_I be the joint law of $X_{t_1} \dots X_{t_N}$, it is necessary and sufficient that the family (μ_I) be consistent.*

Concretely (but this choice is far from unique) one can take $\Omega = E^T$, with X_t: $\Omega \to E$ the coordinate map of index t, $\mathcal{B}$ being the minimal σ-algebra with respect to which all the X_t are measurable; P is then uniquely determined by the μ_I, $I \subset T$.

3.5. The Case of Discrete Time Processes

If $T = \mathbb{Z}$ (resp. $\mathbb{N}$), set $I_n = [-n,n]$ (resp. $I_n = [0,n]$), and

$$\varphi_n = \varphi_{I_n I_{n-1}}, \quad \mu_n = \mu_{I_n} .$$

An elementary argument shows that the family μ_I, I a finite subset of T, is consistent if and only if $\varphi_n(\mu_n) = \mu_{n-1}$ for $n = 1,2, \dots$.

3.6. Example: Sequence of Independent Random Vectors. Random vectors $X_i: \Omega \to E$, $1 \leqslant i \leqslant n$, with $E = \mathbb{R}^k$ are **independent** if

$$P(X_1 \in A_1, \ldots, X_n \in A_n) = P(X_1 \in A_1) \times \ldots \times P(X_n \in A_n)$$

for every family $A_1 \ldots A_n$ of Borel subsets of E. This is equivalent to

$$E[f_1(X_1) \ldots f_n(X_n)] = E[f_1(X_1)] \times \ldots \times E[f_n(X_n)]$$

for every family of functions $f_j: E \to \mathbb{C}$ such that one of the two sides is well defined. In this situation the distributions π_j of the X_j an the distribution π of $(X_1 \ldots X_n)$ are linked by

(1)
$$\pi(A_1 \times \ldots \times A_n) = \pi_1(A_1) \times \ldots \times \pi_n(A_n)$$
$$\int_{E^n} f_1(x_1) \ldots f_n(x_n) d\pi(x_1 \ldots x_n) = \left[\int_E f_1(x) d\pi_1(x)\right] \times \ldots \times \left[\int_E f_n(x) d\pi_n(x)\right]$$

with A_j, f_j as above.

If the probabilities π_j on E are given a priori there is always a (unique) probability π on E^n verifying (1); π is called the **product distribution** of $\pi_1 \ldots \pi_n$ and is denoted by $\pi_1 \otimes \ldots \otimes \pi_n$. If the π_j have densities g_j then π has a density g given by $g(x_1 \ldots x_n) = g_1(x_1) \times \ldots \times g_n(x_n)$, where $(x_1 \ldots x_n) \in E^n$.

Now let π_j be an infinite sequence of probabilities on E. Then there exists a sequence of independent random vectors X_j such that the distribution of X_j is π_j. This is deduced from Theorem 1.3.4 by letting

$$\mu_{j_1 \ldots j_n} = \pi_{j_1} \otimes \ldots \otimes \pi_{j_n},$$

which clearly defines a consistent family.

4. Second-Order Process

4.1. L^2-Space

The space $L^2(\Omega,P)$ is the space of equivalent classes of complex random variables $X: \Omega \to \mathbb{C}$ such that

$$\|X\|_2 = [E(|X|^2)]^{1/2}$$

is finite. Endowed with the scalar product $\langle X,Y\rangle = E(X\bar{Y})$ this space is a **complex Hilbert space** (cf. Appendix). We denote by $L^2_{\mathbb{R}}(\Omega,P)$ the (real) subspace of $L^2(\Omega,P)$ formed by the real valued r.v. Recall that $E|X| \leq \|X\|_2$.

For $X,Y \in L^2(\Omega,P)$ denote by $\sigma^2(X) = E[|X - EX|^2]$ the **variance** of X and $\Gamma(X,Y)$ the **covariance** $E[(X - EX)(\overline{Y - EY})]$ of X and Y. The number $\rho(X,Y) = \Gamma(X,Y)/[\sigma(X)\sigma(Y)]$ is the **correlation coefficient** of X and Y. When $\rho(X,Y) = 0$, X and Y are said to be **uncorrelated**; independent variables are always uncorrelated but the converse is generally false.

4.2. Covariance Matrix

The expectation of a random vector or of a matrix with random coefficients is defined coordinate by coordinate. If X is a random vector of dimension k, with coordinates in $L^2(\Omega,P)$, its expectation $E(X)$ is in $\mathbb{C}^k$ and the covariance matrix $\Gamma(X)$ of X is defined by $\Gamma(X) = E[(X - EX)(X - EX)^*]$ where X, EX are *column matrices* and A^* is *the conjugate transpose* of A.

The square matrix $\Gamma(X)$, or order k, is *positive hermitian*, since $\Gamma(X) = \Gamma(X)^*$ and

$$v^*\Gamma(X)v = E|v^*(X - EX)|^2 \geq 0 \quad \text{for } v \in \mathbb{C}^k.$$

In particular $v^*\Gamma(X)v = 0$ is equivalent to $v^*(X - EX) = 0$, P-almost surely.

Thus $\Gamma(X)$ *is positive definite if and only if there is no affine proper subspace F of $\mathbb{C}^k$ such that $X \in F$, P.* a.s.

4.3. Second-Order Processes

We call **second-order process** any random process X, indexed by an arbitrary set T, with values in $\mathbb{C}$, such that $X_t \in L^2(\Omega,P)$ for all $t \in T$. The **mean** m and the **covariance** K of X are the functions $m(t) = E(X_t)$ and $K(s,t) = \Gamma(X_s,X_t)$, $s,t \in T$. We say that X is **centered** if $m(t) \equiv 0$.

For $T = \mathbb{Z}$ or $\mathbb{N}$, X is also called a **time series.** Multidimensional time series, i.e. $\mathbb{R}^k$ or $\mathbb{C}^k$-valued are frequently (and usefully) considered. *We shall limit ourselves here to the study of one-dimensional time series*, essential from the conceptual point of view, since the mathematical tools remain the same for any finite dimension, while the actual computations and estimation techniques are clearly more complex in dimension $k \geqslant 2$.

4.4. Proposition. *If X is a second-order process indexed by T, its covariance K: $T \times T \to \mathbb{C}$ is a "definite nonnegative function"*, i.e., *for all r, all $t_1 \dots t_r \in T$ and $v_1 \dots v_r \in \mathbb{C}$, one has*

$$K(t_i,t_j) = \overline{K(t_j,t_i)}\ , \quad 1 \leqslant i,j \leqslant r$$

and

$$\sum_{1 \leqslant i,j \leqslant r} v_i K(t_i,t_j) \overline{v}_j \geqslant 0.$$

Proof. Notice that the matrix $A = (A_{ij})$, where $A_{ij} = K(t_i,t_j)$, is the covariance matrix of the vector $(X_{t_1} \dots X_{t_r})$. □

Bibliographical Hints

Concerning the basics of probability theory (Section 2) and the practice of probability calculus, we suggest *Billingsley*, *Breiman*, *Dacunha-Castelle and Duflo* (mainly Chapters 1,3,4 in Volume I) and *Neveu*. At a more elementary and more intuitive level, *Hoel*, *Port*, *Stone* and *Ventsel*.

Chapter II
GAUSSIAN PROCESSES

1. The Use (and Misuse) of Gaussian Models

A time series is called gaussian if the joint distribution of every finite family of observations is gaussian. Among second-order processes, the gaussian processes play an essential role.

Indeed, the second-order moments of a gaussian process completely determine the global distribution of the process, which is far from being the case for arbitrary second-order processes. Moreover, in the gaussian case, the notions of noncorrelation and independence coincide, and conditioning by r.v. $X_1 ... X_n$ is achieved by orthogonal projection on the vector space generated by $X_1 ... X_n$. Thus for gaussian processes *the techniques of optimal linear prediction* - the only ones which are easy to implement - *are in fact optimal in a much stronger sense* (cf. Chapter 4).

The fact that the gaussian model allows such comfortable computations has made it extremely popular in theory as well as in practice. One often invoked standard argument for this gaussian ubiquity is the central limit thoerem. This result insures that the sum of large numbers of independent random variables with bounded variances, has, after adequate normalization, an asymptotically gaussian distribution. From this to the assertion that "every accumulation of roughly independent random shocks is exactly gaussian," there is a tempting epistemological jump, but a jump often lacking a concrete mathematical foundation, as shown for instance by the computations of speeds of convergence in the central limit theorem.

Our point of view is the following; in many concrete cases (specifically in econometrics) available information is too scarce to guarantee that the phenomenon at hand "is" gaussian, or in any case to discover with any reasonable accuracy the law of the observed process. One may then, for lack of any better approach, use techniques of prediction/estimation whose *theoretical* value is very linked to the gaussian model, but one must be content with a totally pragmatic appreciation of the *practical* value of these techniques, without relying on ideal hypotheses which are surely far from reality.

2. Fourier Transform: A Few Basic Facts

Let π be a probability on $\mathbb{R}^k$. **Its Fourier transform** $\hat{\pi}: \mathbb{R}^k \to \mathbb{C}$ is the continuous bounded function

$$\hat{\pi}(z) = \int_{\mathbb{R}^k} e^{i<z,x>} d\pi(x),$$

where $z \in \mathbb{R}^k$, and $<.,.>$ is the usual scalar product on $\mathbb{R}^k$. The function $\hat{\pi}$ determines π uniquely.

Let X be an $\mathbb{R}^k$-valued random vector. The **characteristic function** φ^X of X is the Fourier transform of the distribution π of X so that, for $z \in \mathbb{R}^k$,

$$\varphi^X(z) = \int_{\mathbb{R}^k} e^{i<z,x>} d\pi(x) = E[e^{i<z,X>}].$$

Random vectors $X_1 ... X_n$ with values in $\mathbb{R}^k$ are independent if and only if the vector $Y = (X_1 ... X_n)$ verifies for $z_1, ..., z_n \in \mathbb{R}^k$

$$(1) \qquad \varphi^Y(z_1, ..., z_n) = \varphi^{X_1}(z_1) \times ... \times \varphi^{X_n}(z_n).$$

Indeed, the definition of product measures (cf. Ch. 1) shows that if $\pi_1 ... \pi_n$ are probabilities on $\mathbb{R}^k$ and if $\pi = \pi_1 \otimes ... \otimes \pi_n$, then $\hat{\pi}(z_1, ..., z_n) = \hat{\pi}(z_1) \times ... \times \hat{\pi}(z_n)$ for $z_1 ... z_n \in \mathbb{R}^k$. If $\eta, \pi_1, ..., \pi_n$ are the distributions of $Y, X_1, ..., X_n$, relation (1) is thus equivalent to $\hat{\eta} = \hat{\pi}$, and hence to $\eta \equiv \pi$, that is to the independence of $X_1 ... X_n$.

3. Gaussian Random Vectors

3.1. Definition. For $m \in \mathbb{R}$ and $\sigma > 0$, we call $N(m,\sigma)$ the probability on $\mathbb{R}$ with density

$$f(x) = \frac{1}{\sigma\sqrt{2\pi}}\exp\left[-\frac{(x-m)^2}{2\sigma^2}\right].$$

This is **in dimension 1, the gaussian (or normal) distribution** with mean m and variance σ^2; if the r.v. Y has law $N(m,\sigma)$ one has $E(Y) = m$ and variance $(Y) = \sigma^2$. The characteristic function of Y is given by (cf. [15], p. 75)

$$\varphi^Y(z) = \exp\left[izm - \frac{1}{2}z^2\sigma^2\right] \qquad z \in \mathbb{R} \tag{2}$$

Let us go to dimension $k \geq 1$. A probability π on $\mathbb{R}^k$ is called **gaussian** if for every linear map q: $\mathbb{R}^k \to \mathbb{R}$ the image of π by q is a gaussian distribution. A (real) random vector X will be called gaussian if its distribution is gaussian. Clearly, for such an X, *the mean* $m = EX$ *and the covariance matrix* $\Gamma = \Gamma(X)$ *exist* since the coordinates of X are necessarily gaussian and hence square integrable. Moreover, if $k = \dim X$, for all matrices A of type (p,k) and B of type $(p,1)$, *the vector* $Y = AX + B$ *is again gaussian and one has* $E(Y) = AE(X) + B$, $\Gamma(Y) = A\Gamma(X)A^*$. This is a direct consequence of the definitions.

3.2. Proposition. *For a probability* π *on* $\mathbb{R}^k$ *to be Gaussian, it is necessary and sufficient that its Fourier transform* $\hat{\pi}$ *be of the form*

$$\hat{\pi}(z) = \exp\left[iz^*m - \frac{1}{2}z^*\Gamma z\right] \qquad z \in \mathbb{R}^k \tag{3}$$

with $m \in \mathbb{R}^k$ *and* Γ *nonnegative symmetric matrix of order* k.

Moreover for arbitrary $m \in \mathbb{R}^k$, $\Gamma \geq 0$ *there exists a unique probability* π *on* $\mathbb{R}^k$ *verifying* (3). *The distribution* π *is gaussian, with mean* m *and covariance matrix* Γ; *we denote it by* $N(m,\Gamma)$.

The distribution $N(m,\Gamma)$ *has a density* f *on* $\mathbb{R}^k$ *if and only if* $\det(\Gamma) \neq 0$, *and* f *is then given by*

$$f(x) = (2\pi)^{-k/2}(\det \Gamma)^{-1/2}\exp\left[-\frac{1}{2}(x-m)^*\Gamma^{-1}(x-m)\right], \quad x \in \mathbb{R}^k.$$

Proof. Let π be a gaussian probability on $\mathbb{R}^k$; let Y be a random vector with law π. Set $m = EY$, $\Gamma = \Gamma(Y)$. For z *fixed* in $\mathbb{R}^k$ consider the linear function q: $\mathbb{R}^k \to \mathbb{R}$ given by

$q(x) = z^*x$. By definition $Z = q(Y)$ is gaussian and $E[q(Y)] = z^*m$, $\sigma^2[q(Y)] = z^*\Gamma z$. By (2), the characteristic function of Z is

$$\varphi^Z(u) = \exp\left\{ iuz^*m - \frac{1}{2} u^2 z^*\Gamma z\right), \qquad u \in \mathbb{R}.$$

But the characteristic function of Y verifies

$$\varphi^Y(z) = E[e^{iz^*Y}] = \varphi^Z(1)$$

so that $\hat{\pi}$ is indeed of the form (3). Conversely if π verifies (3) and is the distribution of Y, the same computation proves that for every linear function q, the one-dimensional r.v. $q(Y)$ is gaussian, and hence π is gaussian.

Let $X_1...X_k$ be independent r.v. having the same distribution $N(0,1)$ so that

$$\varphi^{X_j}(u) = \exp\left[-\frac{u^2}{2}\right].$$

By (1) the characteristic function of $X = (X_1...X_k)$ verifies for all $z = (z_1...z_k) \in \mathbb{R}^k$,

$$\varphi^X(z_1...z_k) = \varphi^{X_1}(z_1)...\varphi^{X_k}(z_k) = \exp\left[-\frac{1}{2}z^*z\right].$$

Thus φ^X is of type (3) with $m = 0$, $\Gamma = I$ and X is **standard gaussian** with distribution $N(0,I)$ where I is the identity matrix.

Start with given arbitrary $m \in \mathbb{R}^k$ and Γ symmetric nonnegative, of order k. By diagonalization there is an $A = \sqrt{\Gamma}$ such that $A = A^*$ and $A^2 = \Gamma$. The random vector $Y = AX + m$ is gaussian with mean m and covariance $AIA^* = \Gamma$. Hence the existence of the distribution $N(m,\Gamma)$.

The X_j are independent, with the same density

$$f(v) = \frac{1}{\sqrt{2\pi}} \exp\left[-\frac{v^2}{2}\right].$$

Hence the distribution $N(0,I)$ of X has the density $g(x)$, $x \in \mathbb{R}^k$, given by

$$g(x) = g(x_1...x_k) = f(x_1)...f(x_k) = (2\pi)^{-k/2}\exp\left[-\frac{1}{2}\, x^*x\right].$$

Since $\Gamma = A^2$, $\det \Gamma = (\det A)^2$ and A is invertible if and only if $\det \Gamma \neq 0$. In this case the formula for change of variables in multiple integrals (cf. [14], p. 132) shows that the law $N(m,\Gamma)$ of $Y = AX + m$ must have a density $h(y)$, $y \in \mathbb{R}^k$,

$$h(y) = \frac{1}{|\det A|} g[A^{-1}(y - m)]$$

$$= (2\pi)^{-k/2}(\det \Gamma)^{-1/2}\exp\left[-\frac{1}{2}(y-m)^*\Gamma^{-1}(y-m)\right].$$

If $\det \Gamma = 0$, A has rank $r < k$, with probability one, $Y = AX + m$ takes its values in an affine strict subspace of $\mathbb{R}^k$, namely $F = A(\mathbb{R}^k) + m$ which has dimension r. Since F has measure 1 for $N(m,\Gamma)$ and zero measure for Lebesgue measure on $\mathbb{R}^k$, $N(m,\Gamma)$ cannot have a density. □

3.3. Independence and Zero Correlation

If X, Y are independent k-dimensional real random vectors, having second order moments, call $\Gamma(X,Y) = E(X - EX)(Y - EY)^*]$ where X, EX, Y, EY are considered as column vectors, **the covariance matrix of X and Y.** The independence of X and Y always implies $\Gamma(X,Y) = 0$; the converse is generally false, but is true *if the joint distribution of X and Y is gaussian.*

Proposition. *Let $X_1...X_r$ be $\mathbb{R}^k$-valued random vectors. Assume the joint distribution of $X_1...X_r$ to be gaussian. Then the $X_1...X_r$ are independent if and only if the matrices $\Gamma(X_i,X_j)$, $i \neq j$, $1 \leq i,j \leq r$ are equal to zero.*

Proof. To simplify let us consider only the case $k = 1$. Assume the $\Gamma_{ij} = \Gamma(X_i,X_j)$ to be zero for $i \neq j$. For $z = (z_1...z_r) \in \mathbb{R}^r$ one has, with the notation $X = (X_1...X_r)$,

$$z^*\Gamma(X)z = \sum_{j=1}^{r} z_j^2 \Gamma_{jj}.$$

From (3) the characteristic functions of $X, X_1...X_r$ then verify

$$\varphi^X(z) = \varphi^{X_1}(z_1) \ldots \varphi^{X_r}(z_r)$$

and the X_j are independent (cf. Section 2). The converse is obvious. □

Corollary. *Let $X_1...X_r$, Y be real random variables, with zero means, defined on (Ω,P). Assume the joint law of $(X_1...X_r,Y)$ to be gaussian. Then one has $E(Y|X_1...X_r) = Z$ where Z is the orthogonal projection (in $L^2_{\mathbb{R}}(\Omega,P)$) of Y on the vector space* *generated by* $X_1 \ldots X_r$.

Proof. Write $X = (X_1 \ldots X_r)$. By definition (cf. [15], p. 243) the r.v. $V = E(Y|X)$ is of the form $g(X)$ where $g: \mathbb{R}^r \to \mathbb{R}$ is a measurable function and is uniquely determined in $L^2_{\mathbb{R}}(\Omega,P)$ by the relation

$$E[g(X)f(X)] = E[Yf(X)]$$

where $f: \mathbb{R}^r \to \mathbb{R}$ is an arbitrary bounded measurable function. Let Z be the projection of Y on the vector space generated by $X_1 \ldots X_r$. One has $Z = a_1 X_1 + \ldots + a_r X_r$ and the random vector $(Y - Z, X)$ obtained by linear transformation from the gaussian vector (Y,X) is again gaussian.

By definition of Z, we have $\Gamma(Y - Z, X) = 0$ and the preceding proposition implies that $Y - Z$ and X are independent. In particular for f as above we have

$$E[(Y - Z)f(X)] = E(Y - Z)E[f(X)].$$

The hypothesis $EY = 0$ and $EX = 0$ implies $E(Y - Z) = 0$ and hence

$$E[Zf(X)] = E[Yf(X)].$$

Consequently $V = Z$, P-a.s., and $g(X) = a_1 X_1 + \ldots + a_r X_r$. □

4. Gaussian Processes

4.1. Definition. A random process X indexed by T is called gaussian if for every finite subset $t_1 \ldots t_r$ of T the joint distribution of $X_{t_1} \ldots X_{t_r}$ is gaussian.

4.2. Proposition. *Let* $m: T \to \mathbb{R}$ *be an arbitrary function, and let* $K: T \times T \to \mathbb{R}$ *be a nonnegative definite function* (cf. 1.4.4.). *Then there is a real gaussian process* X *indexed by* T, *with mean* m *and covariance* K. *This process is unique up to equivalence.*

Proof. To every finite subset $I = \{t_1 \ldots t_r\}$ of T, associate the vector $m_I = (m(t_1) \ldots m(t_r))$ in $\mathbb{R}^r$ and the square matrix Γ_I with general term $\Gamma_{ij} = K(t_i, t_j)$. By hypothesis, Γ_I is symmetric nonnegative. Let π_I be the gaussian distribution $N(m_I, \Gamma_I)$.

The family π_I, I finite subset of T, is clearly consistent and Theorem 1.4.4 grants the existence of a process X indexed by T, whose joint distributions are the π_I. The functions m, K determine the π_I, and hence the equivalence class of X, which is of course gaussian by construction.

For the following statements, it will be convenient to consider *any almost surely constant* r.v. *as a gaussian variable with zero variance.* Thus we shall avoid the terms "gaussian or constant" when speaking about limits of gaussian r.v.

4.3. Proposition. *Let X be an $\mathbb{R}$-valued gaussian process indexed by T. Let $H_{\mathbb{R}}^X$ be the smallest closed vector subspace of $L_{\mathbb{R}}^2(\Omega,P)$ containing all the X_t, $t \in T$. Then for all $Z_1, \ldots, Z_r \in H_{\mathbb{R}}^X$ the joint law of $(Z_1 \ldots Z_r)$ is gaussian. In particular two finite families $(Z_1 \ldots Z_r)$ and $(U_1 \ldots U_s)$ of elements of $H_{\mathbb{R}}^X$ are independent of each other if and only if the covariances $\Gamma(Z_i,U_j)$ are zero for all $i \in [1,r]$, $j \in [1,s]$.*

Proof. Let $V_n = \Omega \to \mathbb{R}^k$ be a sequence of gaussian random vectors and V be a r.v. such that $\lim_{n\to+\infty} E\|V_n - V\|^2 = 0$, where $\|\ \|$ is the euclidean norm on $\mathbb{R}^k$. Let us show that V is then gaussian.

One first checks that the mean m_n and covariance Γ_n of V_n converge toward the mean m and covariance Γ of V. Hence the characteristic functions of the V_n verify, for $z \in \mathbb{R}^k$,

$$\lim_{n\to+\infty} \varphi^{V_n}(z) = \lim_{n\to+\infty} \exp\left[iz^*m_n - \frac{1}{2}\, z^*\Gamma_n z\right]$$

$$= \exp\left[iz^*m - \frac{1}{2}\, z^*\Gamma z\right].$$

The L^2-convergence implies the almost sure convergence of a subsequence, so that there exists (n_j) such that

$$\lim_{j\to+\infty} V_{n_j} = V,$$

P-a.s. By the dominated convergence theorem we conclude that

$$\varphi^V(z) = E(\exp iz^*V) = \lim_{j\to+\infty} E(\exp iz^*V_{n_j})$$

$$= \lim_{j\to+\infty} \varphi^{V_{n_j}}(z)$$

and the expression of $\varphi^V(z)$ thus obtained shows that V is gaussian.

To return to Proposition 4.3, the space $H_{\mathbb{R}}^X$ is the closure in $L_{\mathbb{R}}^2$ of the space F, set of all linear combinations $(c_1X_{t_1} + \ldots + c_pX_{t_p})$ where the $t_j \in T$, the $c_j \in \mathbb{R}$ and the integer p are arbitrary. Every finite family of elements of F is a gaussian random vector, being the image of a gaussian vector by a linear map.

If the coordinates of a random vector V belong to $H_{\mathbb{R}}^X$, V is then a limit in $L_{\mathbb{R}}^2$ of random vectors V_n whose coordinates are r.v. belonging to F. The V_n are then gaussian, and so is their limit V. The last assertion in 4.3 is then a consequence of Proposition 3.3.

Note that the elements of $H_{\mathbb{R}}^X$ may be constants, for instance if they are limits of elements whose variance tend to zero. If $EX_t = 0$ for all $t \in T$, then the only constant which may belong to $H_{\mathbb{R}}^X$ is the zero r.v.

Bibliographical Hints

The basic notions on gaussian r.v., the computations of characteristic functions and conditional expectations may be found in *Billingsley*, *Breiman*, and *Dacunha-Castelle-Duflo*, Chs. 5 and 6, Vol. I. The theory of general gaussian processes is given in *Neveu*.

Chapter III
STATIONARY PROCESSES

1. Stationarity and Model Building

Numerous concrete phenomena, in meteorolgy for instance (daily temperatures, monthly rain levels, etc.) or in econometrics (stock exchange rates, price index, unemployment level,...) provide numerical sequences of observations x_n having two characteristics:

(a) besides a more or less obvious global trend (roughly polynomial evolutions, or periodic in n), the graph of the x_n presents a very irregular local aspect, impossible to modelize by a simple curve depending on a *small* number of parameters;

(b) it is impossible to start the sequence of measurements again in "*identical*" conditions.

In view of (a), it is tempting to modelize the sequence x_n by a random process X, which is equivalent to asserting that (x_n) is a "typical" trajectory of the process (X_n). By (b), to be able to deduce the model from the data, the sequence (x_n), $1 \leqslant n \leqslant N$, must determine, at least when $N \to +\infty$, the joint distributions of the process X. One must hence restrict oneself to processes $X = \{\Omega, P, (X_n)\}$ such that for P-almost all $\omega \in \Omega$, every sufficiently long portion of the trajectory $X_n(\omega)$, $1 \leqslant n \leqslant N$, provides sufficiently accurate indications about the distribution P.

The random processes whose probabilistic structure is invariant by time translations (*strict stationarity*) and whose trajectories are sufficiently unpredictable (*ergodicity*) have this property (cf. Section 3.2). Sequences of independent identically distributed random variables are a good example of such processes.

Since we are trying to modelize phenomena which are a priori nonstationary, and verify (a), (b), it seems reasonable to restrict ourselves to modelizing by random processes Y of the following type:

(1) $$Y_n = F(n, X_n, X_{n-1}, ..., X_{n-k}, ...)$$

where X is a *strictly stationary ergodic process* and where $F: \mathbb{Z} \times \mathbb{R}^{\mathbf{N}} \to \mathbb{R}$ is a *deterministic* function, *invertible* in the following sense: the data $(n, Y_n, Y_{n-1}, ...)$ determine completely $(n, X_n, X_{n-1}, ...)$. We have restricted $F(n,X)$ to depend on the *past* of X at time n only, with the same restriction on $F^{-1}(n,Y)$ to preserve a notion of causality, useful when the problem is to predict future values of Y given the past of Y.

The most commonly used models, of the type "ARMA with deterministic trend" (cf. Ch. 9) are of the form

$$Y_n = f(n) + \sum_{k=0}^{+\infty} a_k X_{n-k}$$

where the trend $f(n)$ is deterministic, the a_k are fixed numbers, and X is a sequence of independent identically distributed r.v. Models of the type

$$Y_n = f(n) + \sum_{k=0}^{n} \psi(k) X_{n-k}$$

with f and ψ deterministic, satisfying finite recursions, and X as above, are also often used (but not in this form!); for instance the "ARIMA" models, the "ARMA models with seasonal effects" (cf. Ch. 9) belong to this category.

We are essentially going to study the cornerstone of models of type (1): the strictly stationary ergodic processes.

2. Strict Stationarity and Second-Order Stationarity

Let X be a random process indexed by $T = \mathbb{N}$ or $\mathbb{Z}$. One says that X is **strictly stationary** if for every finite family of instants $t_1 ... t_k \in T$ and every integer r the joint distributions of $(X_{t_1} ... X_{t_k})$ and of $(X_{r+t_1} ... X_{r+t_k})$ are the same.

One says that X is **second-order stationary** (or again **stationary in the wide sense**), if X is a second order process whose mean $M(t)$ and covariance $K(s,t)$ are invariant by time translations, and hence of the form, with $s,t \in T$

$$E(X_t) = M(t) \equiv M \quad \text{and} \quad \Gamma(X_s, X_t) = K(s,t) = \gamma(t-s)$$

where M is a constant and $\gamma: \mathbb{Z} \to \mathbb{C}$ a function such that $\gamma(n) = \overline{\gamma(-n)}$.

Clearly, for a second-order process, **strict stationarity implies second-order stationarity.**

But there are second-order stationary processes which are not strictly stationary; for instance a sequence Y of independent r.v. with identical means and identical variances is always second-order stationary, but Y is strictly stationary if and only if the Y_n have the same distribution.

In the gaussian case, the two notions of stationarity coincide.

Proposition. *A gaussian process X indexed by $\mathbb{N}$ or $\mathbb{Z}$ is strictly stationary if and only if it is second order stationary.*

Proof. If X is gaussian, the joint distributions of $(X_{t_1} ... X_{t_k})$ and $(X_{r+t_1} ... X_{r+t_k})$ are gaussian; they coincide if and only if they have the same mean and the same covariance, whence the result.

Second-order stationarity is much more accesible to statistical verification than strict stationarity. Its practical importance is especially linked to problems of prediction or regression. Indeed to deal with effectively computable estimators, it is common practice to use **linear** optimal predictors (cf. Ch. 4) whose computation does not involve the detailed probabilistic structure of the observed X process, but only the "geometry" (angles and lengths) of the sequence X_n

considered as a sequence of vectors in the Hilbert space $L^2(\Omega,P)$, and this "geometry" only depends on the covariances $\Gamma(X_m,X_n)$ and the means $E(X_n)$ of X. From this point of view the natural notion of stationarity is the invariance of second-order moments by time translations.

3. Construction of Strictly Stationary Processes

3.1. Filters and Stationarity

If X is a strictly stationary time series, with values in $\mathbb{R}$, and if F: $\mathbb{R}^{k+r+1} \to \mathbb{R}$ is an arbitrary measurable function, *the process* $Y_n = F(X_{n-k}, X_{n-k+1}, \ldots, X_{n+r})$ *is again strictly stationary.* For instance, starting with a gaussian stationary X, one builds in this way an infinity of strictly stationary processes, generally nongaussian. This extends to functions F: $\mathbb{R}^{\mathbb{Z}} \to \mathbb{R}$. The operation $Y = F(X)$ is an example of **nonlinear filter.** But if there is no strict stationarity for X, *then second-order stationarity is generally not preserved by a filter F of the above type, except if F is linear.* The crucial case of linear filters will be studied in Ch. 6.

3.2. The Past of a Strictly Stationarity Process

Let $X = (X_n)_{n \geq 0}$ be a strictly stationary process given for $n \geq 0$ only. We want to build a past for X without losing its strict stationarity. The problem is to find a strictly stationary Y indexed by $\mathbb{Z}$ such that $(X_n)_{n\geq 0}$ and $(Y_n)_{n\geq 0}$ have the same law on $\mathbb{R}^{\mathbb{N}}$.

Proposition. *For every strictly stationary process* X_n, $n \geq 0$, *there is a strictly stationary process* Y_n *index by* $n \in \mathbb{Z}$, *unique up to equivalence, such that* $(X_n)_{n\geq 0}$ *and* $(Y_n)_{n\geq 0}$ *are equivalent processes.*

Proof. Let $I = \{n_1 \ldots n_k\} \subset \mathbb{Z}$. For all r large enough, the integers $r+n_1, \ldots, r+n_k$ are in $\mathbb{N}$ and the joint law of $X_{r+n_1} \ldots X_{r+n_k}$ is independent of r, since X is strictly stationarity. Let π_I be this law. The family π_I being obviously consistent, Kolmogorov's Theorem 1.3.4 grants the existence of a process Y indexed by $\mathbb{Z}$ such that for all I as above, π_I is the joint law of $(Y_{n_1} \dots Y_{n_k})$. Clearly Y has the announced properties.

4. Ergodicity

We sketch the basic definitions and theorems, referring to [8] for proofs and useful complements. The mathematical content of this paragraph will not be used in the sequel, which should not slight its importance from a conceptual point of view.

Let X be a strictly stationary random process indexed by $\mathbb{Z}$, with values in $\mathbb{R}$. Assume that the X_n are defined on a probability space (Ω, P). To X we associate a probability Q on $\mathbb{R}^{\mathbb{Z}}$, the **law of the trajectories** of X, where $\mathbb{R}^{\mathbb{Z}}$ is endowed with the smallest σ-algebra for which all coordinate maps become measurable. More precisely, if $u: \Omega \to \mathbb{R}^{\mathbb{Z}}$ is defined by $u(\omega) = \{X_n(\omega)\}_{n \in \mathbb{Z}}$, then Q is the image of P by u.

On the other hand let $\theta: \mathbb{R}^{\mathbb{Z}} \to \mathbb{R}^{\mathbb{Z}}$ be the time translation defined by

$$\theta[(x_n)_{n \in \mathbb{Z}}] = (y_n)_{n \in \mathbb{Z}} \quad \text{with} \quad y_n \equiv x_{n+1}.$$

Then X is strictly stationary if and only if the law Q is invariant by θ, that is if $Q(\theta^{-1}(A)) = Q(A)$ for every measurable subset A of $\mathbb{R}^{\mathbb{Z}}$.

4.1. Definition. A strictly stationary process X is called ergodic if the measurable subsets A of $\mathbb{R}^{\mathbb{Z}}$ invariant by θ are necessarily of Q-measure zero or one, that is if $\theta^{-1}(A) = A$ implies $Q(A) = 0$ or $Q(A) = 1$.

Intuitively this means that almost every trajectory of X completely fills, by successive translations, the whole space $\mathbb{R}^{\mathbb{Z}}$. This implies a very strong irregularity for the trajectories of X since for "each" of these trajectories, the future holds any preassigned type of behaviour in reserve (provided it's not a priori eliminated by the law Q). The impartiality of the trajectories implies a time average property, discovered by Birkhoff and von Neumann.

4.2. Ergodic Theorem. *Let X be ergodic strictly stationary. Then if* $g\colon \mathbb{R}^{\mathbb{Z}} \to \mathbb{R}$ *is a Q-integrable function, one has for Q-almost all* $x \in \mathbb{R}^{\mathbb{Z}}$,

$$\lim_{N\to+\infty} \frac{1}{N}\left[g(x) + g \circ \theta(x) + \dots + g \circ \theta^N(x)\right] = \int_{\mathbb{R}^{\mathbb{Z}}} g\,dQ.$$

More concretely, take g of the following form

$$g(x) = f(x_1 \dots x_k) \quad \textit{for} \quad x = (x_n)_{n\in\mathbb{Z}} \in \mathbb{R}^{\mathbb{Z}}$$

with $f\colon \mathbb{R}^k \to \mathbb{R}$ *such that* $E[f(X_1 \dots X_k)]$ *exists. Then* 4.2 *implies*

$$\text{(2)} \qquad \lim_{N\to+\infty} \frac{1}{N}\left[f(X_1 \dots X_k) + f(X_2 \dots X_{k+1}) + \dots + f(X_{N+1} \dots X_{N+k})\right] = E[f(X_1 \dots X_k)]$$

where the convergence holds P-a.s.

In particular for $f = 1_A$ and $A \subset \mathbb{R}^k$, (2) implies the P.-a.s. convergence

$$\text{(3)} \quad P[(X_1 \dots X_k) \in A] = \lim_{N\to+\infty} \frac{1}{N}[1_A(X_1 \dots X_k) + \dots + 1_A(X_{N+1} \dots X_{N+k})]$$

and *the joint distribution* of $X_1 \dots X_k$ *is hence completely determined as* $N \to +\infty$ *by the sequence of observations* $X_1 \dots X_N$. *We shall say that strictly stationary ergodic processes are identifiable by an infinite trajectory.*

Note that the validity of (2) for every choice of f and k, or even the validity of (3) for every choice of A and k, are **equivalent** to the ergodicity of the strictly stationary process X.

4.3. Theorem ([8], [40]). *Every sequence* X_n, $n \in \mathbb{Z}$ *of independent identically distributed random variables is a strictly stationary ergodic process.*

4.4. Theorem ([46], p. 163). *Every stationary gaussian process X such that*

$$\lim_{|n-m|\to+\infty} \Gamma(X_n, X_m) = 0$$

is ergodic.

5. Second-Order Stationarity: Processes with Countable Spectrum

Let A_k, $k \geqslant 1$ be a sequence of centered complex r.v. belonging to $L^2(\Omega,P)$, pairwise uncorrelated, and with variance $\sigma_k^2 = \sigma^2(A_k)$ such that $\Sigma_{k\geqslant 1}\sigma_k^2 < \infty$. Let $\lambda_k \in [-\pi,\pi[$ be an arbitrary sequence of distinct numbers. Then **the series**

$$X_n = \sum_{k\geqslant 1} A_k e^{in\lambda_k}$$

converges in L^2 and the process $X = (X_n)_{n\in\mathbf{Z}}$ is centered, second-order stationary. An immediate computation indeed gives $EX_n = 0$ and

$$K(m,n) = E(X_m\overline{X}_n) = \sum_{k\geqslant 1} \sigma_k^2 e^{i(m-n)\lambda_k} = \gamma(m-n).$$

A fortiori each process $n \to A_k e^{in\lambda_k}$ is itself second-order stationary, and is, by the way, **a periodic function of the time** n, with period $2\pi/\lambda_k$, hence **with deterministic frequency λ_k but random amplitude A_k**. The process X appears as a linear combination of pairwise uncorrelated processes with deterministic frequencies and random amplitudes. In the terminology borrowed from the theory of hertzian signals or from optics, $\sigma_k^2 = E|A_k|^2$ is called the **energy of the frequency** λ_k. It is remarkable (and nontrivial, cf. Chapter 6) that every second-order stationary process is a limit in L^2 of processes of the type just described.

We shall call this class of processes the family of **processes with countable spectrum**; we shall say that X has **finite spectrum** if the A_k are zero for $k \geqslant N$.

Bibliographic Hints

The notion of ergodicity is the center of this chapter. Besides the book quoted here (*Billingsley*) more information about gaussian processes may be found in *Rozanov* [46].

Chapter IV
FORECASTING AND STATIONARITY

1. Linear and Nonlinear Forecasting

Let (X_t) be a complex second order process defined on (Ω,P). Let $Y \in L^2(\Omega,P)$. To forecast the value of Y after having observed $X_{t_1} \ldots X_{t_k}$ is equivalent to finding a deterministic function $f: \mathbb{C}^k \to \mathbb{C}$ and to forecast Y by $\tilde{Y} = f(X_{t_1} \ldots X_{t_k})$. The **optimal prediction (in the least squares sense)** consists in choosing f such that $E[|Y - \tilde{Y}|^2]$ is minimal. This classical problem admits *a unique solution*

$$\tilde{Y} = E(Y \mid X_{t_1} \ldots X_{t_k}) = \varphi(X_{t_1} \ldots X_{t_k}),$$

the conditional expectation of Y given $X_{t_1} \ldots X_{t_k}$ ([15], p. 242).

In practice, the function φ is often impossible to determine, either because the joint distribution of $(X_{t_1} \ldots X_{t_k}, Y)$ is only roughly known, or because it does not lend itself to explicit computations. This leads to restricting the class of estimators considered and using **linear optimal prediction**: one still seeks to minimize $E[|\hat{Y} - Y|^2]$ with

$$\hat{Y} = f(X_{t_1} \ldots X_{t_k}),$$

but with the restriction f **linear.** The **unique solution** is obviously given by $\hat{Y} = p(Y)$ where p is the orthogonal projection on the vector space generated by $X_{t_1} \dots X_{t_k}$ in $L^2(\Omega,P)$.

In general $E[|Y - \hat{Y}|^2] > E[|Y - \tilde{Y}|^2]$ but in the case where the joint distribution of $(X_{t_1} \dots X_{t_k}, Y)$ is gaussian these numbers coincide since $\hat{Y} = \tilde{Y}$ (cf. Chapter 2).

The point of view of linear prediction, widely prevalent in practice, boils down to considering that the information supplied by the observation of the process $X = (X_t)_{t \in T}$ depends only on *the space* H^X, *linear closed subspace of* $L^2(\Omega,P)$ *generated* by the X_t, $t \in T$, space which we shall call **the linear envelope** of X. The X_t once observed, this is equivalent to considering, as the only r.v. accessible to computation, the finite linear combinations of the X_t and the L^2-limits of such linear combinations.

The space H^X is by construction a *complex* Hilbert space. When X is real we also consider the space $H^X_{\mathbb{R}}$ generated in $L^2_{\mathbb{R}}(\Omega,P)$ by the X_t, $t \in T$. Note that *if* X *is centered one has* $E(V) = 0$ *for all* $V \in H^X$; indeed the map $V \to E(V)$ is linear and continuous on L^2 since $|E(V)| \leqslant \|V\|_2$ so that the set of $V \in L^2(\Omega,P)$ verifying $E(V) = 0$ is a closed vector subspace of L^2.

For $Y \in L^2(\Omega,P)$ the optimal linear forecast of Y given X is hence the orthogonal projection on Y on H^X.

2. Regular Processes and Singular Processes

Let X be a second-order process indexed by $\mathbb{Z}$ defined on (Ω,P). The information associated to the X_k, $k \leqslant n$ is described by the linear envelope H^X_n of the process $(X_k)_{k \leqslant n}$. We call H^X_n **the linear past of** X **at time** n. The spaces H^X_n increase with n. We define the **asymptotic past** of X by

$$H^X_{-\infty} = \bigcap_{n \in \mathbb{Z}} H^X_n,$$

and we write $H^X_{+\infty} = H^X$. The problems of optimal linear forecast given the past of X at time n are solved by orthogonal projection on H^X_n. This projection denoted by p_n: $L^2(\Omega,P) \to H^X_n$ is also known as the **linear predictor at time** n.

The process X will be called **regular** if $H^X_{-\infty} = \{0\}$ and **singular** if $H^X_{-\infty} = H^X$. The only process which is at once regular and singular is the process identical to zero.

2.1. Theorem (Wold). *Every second-order time series X may be written $X = X^r + X^s$ where X^r is a regular process, X^s a singular process, and the linear envelopes of X^r and X^s are orthogonal.*

Proof. For $V \in L^2(\Omega,P)$, the $v_n = p_{n+1}(V) - p_n(V)$ form an orthogonal sequence whence

$$\sum_{-N\leqslant n\leqslant -1} \|v_n\|^2 = \Big\|\sum_{-N\leqslant n\leqslant -1} v_n\Big\|^2$$

$$= \|p_0(V) - p_{-N}(V)\|^2 \leqslant 4\|V\|^2$$

and the series $\Sigma_{n\leqslant -1} v_n$ converge in L^2. Since

$$\sum_{-N\leqslant n\leqslant -1} v_n = p_0(V) - p_{-N}(V)$$

we obtain the existence of $v = \lim_{n\to-\infty} p_n(V)$. As $p_n p_{-\infty} = p_{-\infty} p_n = p_{-\infty}$, we conclude that $v = p_{-\infty}(v)$ and $p_{-\infty}(V - v) = 0$ whence $v = p_{-\infty}(V)$ and

(1) $$\lim_{n\to-\infty} p_n(V) = p_{-\infty}(V) \qquad \text{for all } V \in L^2(\Omega,P).$$

Write $X_n^s = p_{-\infty}(X_n)$ and $X_n^r = X_n - X_n^s$. Since $p_{-\infty}$ is a projection one has $\langle X_m^r, X_n^s\rangle = 0$ for all $m,n \in \mathbb{Z}$, and H^{X_r} is orthogonal to H^{X_s}. The space $H_{-\infty}^X = p_{-\infty}(H_n^X)$ is generated by the $p_{-\infty}(X_k)$, $k \leqslant n$ and hence coincides with $H_n^{X_s}$ for all n. Thus X^s is singular.

Let $V \in H_{-\infty}^{X_r}$; then $V \in H_n^{X_r}$ for all n, and hence is a limit of linear combinations of the $[X_k - p_{-\infty}(X_k)]$ for $k \leqslant n$, which forces $[p_n - p_{-\infty}](V) = V$ for all n. When $n \to -\infty$ we conclude by (1) that $V = 0$. Whence the regularity of X^r.

2.2. White Noise

One calls **white noise** any second-order stationary process W, centered, such that W_n and W_m are uncorrelated for $n \neq m$.

For instance, a sequence of independent centered r.v. having the same variance is a white noise. It is easy to see that *a white noise is always regular.*

2.3. Processes with Finite Spectrum

Let

$$X_n = \sum_{k=1}^{r} A_k e^{in\lambda_k}$$

be a *second-order stationary random process, with finite spectrum* (cf. Chapter 3, Section 5). Let us show that *X is singular.* Write

$$\alpha_{pq}(n) = e^{i(p-n)\lambda_q}, \quad 1 \leqslant p,q \leqslant r.$$

The matrix $\alpha(n)$ with elements $\alpha_{pq}(n)$ is invertible (Van der Monde determinant). Let $\beta(n) = [\beta_{pq}(n)]$ be its inverse. We have

$$A_p = \sum_{q=1}^{r} \beta_{pq}(n) X_{n-q}$$

and hence H_n^X coincide for every n with the space F generated by the A_p, $1 \leqslant p \leqslant r$, whence the result.

3. Regular Stationary Processes and Innovation

3.1. Isometries

Let H_1,H_2 be two complex Hilbert spaces and G be a vector subspace of H_1. An **isometry** S from G into H_2 is a linear map $S: G \to H_2$ such that $\langle Sv,Sw \rangle_{H_2} = \langle v,w \rangle_{H_1}$ for $v,w \in G$.

It is easy to construct isometries with the help of the following technical result, proved in Section 5 below.

Lemma. *Let* $v_t \in H_1$, $w_t \in H_2$ *be two families of vectors indexed by* $t \in T$, *where* T *is an arbitrary set. Assume that*

$$\langle v_s,v_t \rangle_{H_1} = \langle w_s,w_t \rangle_{H_2} \quad \text{for} \quad s,t \in T.$$

Let V and W be the closed vector subspaces of H_1 and H_2 respectively generated by the $(v_t)_{t \in T}$ and the $(w_t)_{t \in T}$. Then there exists a unique isometry S from V into H_2 such that $Sv_t = w_t$ for all $t \in T$. Moreover one has $S(V) = W$.

3.2. The Backward Shift of a Stationary Process

Let X be the second-order stationary process, defined on (Ω,P), and let H^X be its linear envelope. There exists by 3.1 a unique isometry S^X from H^X into $L^2(\Omega,P)$ such that $S^X(X_n) = X_{n-1}$ for all $n \in \mathbb{Z}$. We have $S^X(H^X) = H^X$ and S^X is called **the backward shift** of X.

3.3. Wold's Decomposition and Stationarity

Let X be a centered second order stationary process. Then one has $X = X^r + X^s$ where the regular process X^r and the singular process X^s are centered **stationary** and the linear envelopes of X^r and X^s are orthogonal.

Indeed, let p_n be the projection on H_n^X and S be the backward shift of X. By construction one has

$$(2) \qquad p_n S = S p_{n+1} \qquad n \in \mathbb{Z}$$

whence by (1), $p_{-\infty}S = Sp_{-\infty}$. The Wold decomposition of X is given by $X_n^r = X_n - p_{-\infty}(X_n)$ and $X_n^s = p_{-\infty}(X_n)$ which then implies $SX_n^r = X_{n-1}^r$, $SX_n^s = X_{n-1}^s$. Since S is an isometry, X^r and X^s are second-order stationary.

3.4. Theorem. *Let X be a centered, second-order stationary process. Then X is regular if and only if there exists a white noise W such that $H_n^X = H_n^W$ for all $n \in \mathbb{Z}$. Such a white noise is unique up to multiplication by a constant. In particular one may choose*

$$(3) \qquad W_n = X_n - p_{n-1}(X_n)$$

where p_n is the orthogonal projection on H_n^X. Moreover there exist numbers $c_k \in \mathbb{C}$ such that $\Sigma_{k\geq 0}|c_k|^2 < \infty$ and

(4) $$X_n = \sum_{k \geq 0} c_k W_{n-k} \quad \textit{for all} \quad n \in \mathbb{Z}.$$

Proof. If there is a white noise W such that $H_n^X = H_n^W$ for all n, one has $H_n^X = \mathbb{C}W_n \oplus H_{n-1}^W$ whence

(5) $$H_n^X = \mathbb{C}W_n \oplus H_{n-1}^X \quad \text{for all} \quad n \in \mathbb{Z}$$

and the uniqueness of W up to a multiplicative constant. Moreover one has $H_{-\infty}^X = H_{-\infty}^W$ and the regularity of W implies the regularity of X.

Conversely, let X be regular and define a process W by (3). By (2), the backward shift S of X verifies $SW_n = W_{n-1}$, and S being isometric, W is stationary. By definition $W_n \in H_n^X$ and is orthogonal to H_{n-1}^X. Thus W is a white noise and verifies (5) by construction.

By iteration (5) implies for all $j \geq 0$

$$H_n^X = \mathbb{C}W_n \oplus \mathbb{C}W_{n-1} \oplus \ldots \oplus \mathbb{C}W_{n-j} \oplus H_{n-j-1}^X$$

whence the existence of $c_k \in \mathbb{C}$, $0 \leq k \leq j$ such that, n being fixed,

$$X_n = \sum_{k=0}^{j} c_k W_{n-k} + p_{n-j-1}(X_n).$$

Since X is regular, $p_{-\infty}$ is zero and (1) implies

$$\lim_{j \to +\infty} p_{n-j-1}(X_n) = p_{-\infty}(X_n) = 0.$$

For each n we thus obtain numbers c_k, a priori depending on n, such that

$$X_n = \sum_{k=0}^{+\infty} c_k W_{n-k}$$

where the series converges in L^2. The relations $SX_n = X_{n-1}$ and $SW_n = W_{n-1}$ prove that the c_k are independent of n, since (4) determines the $c_k(n)$ for fixed n, due to the orthogonality of the W_j. This orthogonality also implies $\Sigma_{k \geq 0} c_k^2 < \infty$.

The representation (4) gives $X_m \in H_m^W$ for $m \leq n$ and hence $H_n^X \subset H_n^W$. On the other hand (3) implies $W_m \in H_n^X$ for $m \leq n$, whence $H_n^W \subset H_n^X$ and finally $H_n^W \equiv H_n^X$.

3.5. Innovation

Let X be second-order stationary, centered, regular. One calls **innovation** of X the white noise W defined by (3). By 3.4, *the linear pasts* H_n^X *and* H_n^W *of* X *and its innovation* W *coincide at each instant* n. Intuitively, if the X_m, $m \leq n-1$ are known, the added information furnished by observation X_n is represented by the innovation W_n.

Note that **if X is a real gaussian process, the innovation of X is also a real gaussian process** since $H^X = H^W$, $H_{\mathbb{R}}^X = H_{\mathbb{R}}^W$, and the result is a consequence of Chapter 2, Prop. 4.3. In particular, *the innovation* W_n, $n \in \mathbb{Z}$, *of* X *is then a sequence of independent* r.v. *with the same law* $N(0,\sigma^2)$.

In the general case, the variance σ^2 of the innovation of X is, in view of (3), given by

$$\sigma^2 = E[|X_n - p_{n-1}(X_n)|^2].$$

This is the **one step (quadratic) error of prediction** corresponding to the quadratic distance between X_n and its optimal linear forecast $\hat{X}_n = p_{n-1}(X_n)$ based on the whole past of X at time $n-1$.

4. Prediction Based on a Finite Number of Observations

4.1. Linear Regression

Let $Y, V_1, \ldots, V_k$ be real centered square integrable r.v. The **linear regression** $\hat{Y}$ of Y on $(V_1 \ldots V_k)$ is the optimal linear forecast of Y given $V_1 \ldots V_k$. We then have

$$\hat{Y} = \sum_j x_j V_j$$

with $Y - \hat{Y}$ orthogonal to the V_i, whence the system

$$\sum_j E(V_i V_j) x_j = E(V_i Y) = y_i$$

which may be written $\Gamma(V)x = y$ where the covariance matrix $\Gamma(V)$ of $V = (V_1 \ldots V_k)$ is assumed to be *invertible* (i.e. no deterministic linear relationship between the $V_1 \ldots V_k$), to obtain $x = \Gamma(V)^{-1} y$.

The error of prediction is

$$E|Y - \hat{Y}|^2 = E|Y|^2 - E|\hat{Y}|^2 = E|Y|^2 - y^*\Gamma(V)^{-1}y$$

since $x^*\Gamma(V)x = y^*\Gamma(V)^{-1}y$. These computations are easily extended to the complex case.

4.2. Forecasting Given the Limited Past

Let X be a centered, real valued, second-order stationary regular process. **The covariance matrix of a finite number of observations is then always invertible if X is not identically zero.** Indeed if this were not the case, there would be $X_{n_1} \ldots X_{n_r}$ linked by a fixed nontrivial linear relation (cf. Chapter 1, Section 4.2) which would contradict the relation $X_{n+1} \notin H_n^X$, true for every nonzero regular process by 3.4.

Conversely, *without assuming X to be regular*, assume that for every $k \geq 1$, the covariance matrix Γ_k of $(X_{n-1} \ldots X_{n-k})$ which, by the way, coincides with the covariance matrix of $(X_1 \ldots X_k)$, is invertible. Write $\gamma(m - n) = \Gamma(X_m, X_n)$. The computation 4.1 gives the prediction error σ_k^2 corresponding to the linear regression $\hat{X}_{n,k}$ of X_n on $X_{n-1} \ldots X_{n-k}$:

$$\sigma_k^2 = E|X_n - \hat{X}_{n,k}|^2 = \gamma(0) - y_k^* \Gamma_k^{-1} y_k$$

with

$$y_k^* = (\gamma(1) \ldots \gamma(k)) \quad \text{and} \quad \Gamma_k = [\gamma(i - j)]_{1 \leq i,j \leq k} \, .$$

The block decomposition

$$\Gamma_{k+1} = \begin{bmatrix} \gamma(0) & y_k^* \\ y_k & \Gamma_k \end{bmatrix}$$

implies by elementary arguments

$$\det \Gamma_{k+1} = [\gamma(0) - y_k^* \Gamma_k^{-1} y_k] \det \Gamma_k$$

whence the relation

$$\sigma_k^2 = \frac{\det \Gamma_{k+1}}{\det \Gamma_k} \, . \tag{6}$$

On the other hand, call F the vector space generated by $X_{n-1}...X_{n-k}$, and write the orthogonal decomposition $H^X_{n-1} = G \oplus H^X_{n-1-k}$ and $F = G \oplus H$ with $H \subset H^X_{n-1-k}$, which are consequences of the definitions of F, H^X_m. Call f, g, h the projections on F, G, H, so that $f = g + h$, $p_{n-1} = g + p_{n-1-k}$, $h = hp_{n-1-k}$. We get $\hat{X}_{n,k} = fX_n = gX_n + hp_{n-1-k}X_n$ and $p_{n-1}X_n = gX_n + p_{n-1-k}X_n$ whence, denoting by I the identity,

$$X_n - \hat{X}_{n,k} = X_n - p_{n-1}(X_n) + (I - h)p_{n-1-k}(X_n)$$

and consequently with $\sigma^2 = E|X_n - p_{n-1}(X_n)|^2$,

$$\sigma_k^2 = E|X_n - \hat{X}_{n,k}|^2 = \sigma^2 + \|(I - h)p_{n-1-k}(X_n)\|^2.$$

Since $\|I - h\| \leq 1$ and $\lim_{k\to+\infty} p_{n-1-k}(X_n) = p_{-\infty}(X_n) = 0$, the one step error of prediction σ^2 verifies

$$\sigma^2 = \lim_{k\to+\infty} \sigma_k^2 . \tag{7}$$

The limit of a sequence is also the limit of its Cesaro means, which implies, thanks to (6) and (7),

$$\log \sigma^2 = \lim_{k\to+\infty} \frac{1}{k} (\log \sigma_1^2 + ... + \log \sigma_k^2)$$

$$= \lim_{k\to+\infty} \frac{1}{k} \log \det \Gamma_{k+1}.$$

This result may be summarized as follows.

4.3. Proposition. *Let X be a real valued, centered, second-order stationary random process, and let Γ_k be the covariance matrix of $X_1...X_k$. Then one has*

$$\lim_{k\to+\infty} \frac{1}{k} \log \det \Gamma_k = \log \sigma^2$$

where $\sigma^2 = E|X_n - p_{n-1}(X_n)|^2$ is the one step prediction error of X.

Proof. If one of the Γ_k is not invertible, there is a deterministic linear relation

$$\alpha_1 X_1 + ... + \alpha_{k_0} X_{k_0} = 0.$$

All the Γ_k, $k \geq k_0$ then have zero determinants and

$$\lim_{k\to+\infty} \frac{1}{k}\log \det \Gamma_k = -\infty.$$

But if k_0 is minimal, the linear relation just written implies

$$X_{k_0} \in H^X_{k_0-1}$$

and applying the backward shift S^m to the two terms, $X_n \in H^X_{n-1}$ for all n. The process X is then singular and $\sigma^2 = 0$, whence the result in this case. If all the Γ_k are invertible, the result has already been proved in 4.2.

5. *Complements on Isometrics*

5.1. *Lemma*. *Let H_1, H_2 be two Hilbert spaces. Let F be a vector subspace of H_1, not necessarily closed, and let J be an isometry of F into H_2. Then there is a unique isometry S: $\overline{F} \to H_2$ extending J from F to $\overline{F}$, and $S(\overline{F})$ is the closure of $J(F)$ in H_2.*

Proof. Let $v \in \overline{F}$. For every sequence $v_n \in F$ converging to v, Jv_n is a Cauchy sequence in H_2 since (v_n) is a Cauchy sequence and J is an isometry. Hence there is a $w \in H_2$ such that

$$w = \lim_{n\to+\infty} Jv_n.$$

If v'_n is another sequence converging to v, we shall have, by isometry,

$$\lim_{n\to+\infty} \|Jv_n - Jv'_n\| = 0$$

and w does not depend on the chosen sequence. Define then $Sv = w$. The properties of linearity and preservation of scalar products are obviously transmitted by passage to the limit and S; $\overline{F} \to H_2$ is an isometry extending to J. The uniqueness of S is obvious.

By construction $S(\overline{F})$ is included in the closure of $J(F)$. conversely if $w = \lim_{n\to\infty} Jv_n$ with $v_n \in F$, (Jv_n) is a Cauchy sequence, and by isometry, so is (v_n). Let v be the limit of the

v_n; one has $Sv = w$ since S is continuous, and $w \in S(\bar{F})$. Whence the equality of $S(\bar{F})$ and the closure of $J(F)$.

5.2. *Construction of Isometries* (Proof of Lemma 3.1)

Let T be an arbitrary set. Consider two families of vectors $v_t \in H_1$ and $w_t \in H_2$ indexed by T, where H_1, H_2 are Hilbert spaces. Assume that

(8) $$\langle v_s, v_t \rangle_{H_1} = \langle w_s, w_t \rangle_{H_2} \quad \text{for } s,t \in T.$$

Let V, W be the closed vector subspaces generated by the v_t, $t \in T$ and the w_t, $t \in T$.

Call F the vector space of *finite* linear combinations of the v_t, with complex coefficients. For

$$f = \sum_j c_j v_{t_j} \in F$$

write

$$Sf = \sum_j c_j w_{t_j} \,.$$

Then (8) implies

$$\langle Sf, Sg \rangle_{H_2} = \langle f, g \rangle_{H_1} \quad \text{for all } f,g \in F.$$

We conclude that $f \to Sf$ defines *without ambiguity* (although the decomposition $f = \Sigma_j c_j v_{t_j}$ is *not unique* in general) a map S: $F \to H_2$, and that S is an isometry. By 5.1, S extends uniquely to an isometry $\hat{S}: \bar{F} \to H_2$ and $\hat{S}(\bar{F})$ is the closure of $S(F)$. But by construction $\bar{F} = V$ and $S(F)$ is the vector space generated by the w_t, $t \in T$, so that $\hat{S}(V) = W$.

We have thus constructed an isometry $\hat{S}$ from V onto W such that $\hat{S}v_t = w_t$ for all $t \in T$. The uniqueness of $\hat{S}$ is obvious. This proves. 3.1.

Bibliographical Hints

As indicated in the introduction, more information on Hilbert spaces may be found in most books concerned with functional analysis, *Rudin* for instance. A deeper analytic study of problems linked to the regularity of stationary processes, especially gaussian processes can be found in the book by *Ibraguimov and Rozanov.*

Chapter V
RANDOM FIELDS AND STOCHASTIC INTEGRALS

We shall see in Chapter 6 that every stationary process is the Fourier transform of a random measure carried by $\mathbb{T} = [-\pi,+\pi[$. We describe here the uncorrelated random fields which formalize the intuitive notion of random measure. The theory extends easily to more general spaces $\mathbb{T}$.

1. Random Measures with Finite Support

Let $\mathbb{T}$ be the interval $[-\pi,+\pi[$. The (deterministic) measures on $\mathbb{T}$ with complex values and finite support are of the form

$$\nu = \sum_{k=1}^{r} a_k \delta_{\lambda_k}$$

where the λ_k are distinct points of $\mathbb{T}$, $a_k \in \mathbb{C}$, and δ_λ is the unit mass at point λ. For every function $f\colon \mathbb{T} \to \mathbb{C}$ one has

$$\int_{\mathbb{T}} f\, d\nu = \sum_k a_k f(\lambda_k).$$

A natural idea for constructing on $\mathbb{T}$ a "random measure" Z with finite support is to fix in $\mathbb{T}$ a (deterministic) set of distinct points λ_k and to assign to the point λ_k a random "mass" A_k, where the A_k are **complex r.v. belonging to the same space** $L^2(\Omega,P)$. For an arbitrary $f\colon \mathbb{T} \to \mathbb{C}$, one may then define $\int_{\mathbb{T}} f\, dZ$ by

$$Z_f = \int_{\mathbb{T}} f\, dZ = \sum_{k=1}^{r} A_k f(\lambda_k).$$

We impose the following hypothesis: **the A_k are pairwise uncorrelated.** For any two functions f, g from $\mathbb{T}$ to $\mathbb{C}$, the r.v. Z_f and Z_g are in $L^2(\Omega,P)$ and their covariance is, with the notation $\sigma_k^2 = \sigma^2(A_k)$

$$\Gamma(Z_f\,,\,Z_g) = \sum_k \sum_\ell f(\lambda_k)\overline{g(\lambda_\ell)}\,\Gamma(A_k,\,A_\ell)$$
$$= \sum_k \sigma_k^2 f(\lambda_k)\overline{g(\lambda_k)}$$

which may be written

$$\Gamma(Z_f\,,\,Z_g) = \int_{\mathbb{T}} f\bar{g}\, d\mu$$

where μ is the real valued, deterministic, positive measure defined on $\mathbb{T}$ by

$$\mu = \sum_k \sigma_k^2\, \delta_{\lambda_k}\,.$$

Our random measure Z thus defines a deterministic positive measure μ on $\mathbb{T}$ and *a map $f \to Z_f$ from $L^2(\mathbb{T},\mu)$ into $L^2(\Omega,P)$ which is clearly linear and preserves scalar products.*

If one notes Z_A the mass given to $A \subset \mathbb{T}$ by the measure Z, that is $Z_A = Z_{1_A}$, one has

$$\Gamma(Z_A,Z_B) = \int_{\mathbb{T}} 1_A 1_B\, d\mu = \mu(A \cap B)$$

and in particular **the random masses of any two disjoint sets are uncorrelated.**

The family of r.v. Z_A, $A \subset \mathbb{T}$, may be considered as a second-order process, where the set of "times" T is the set of all subsets A of $\mathbb{T}$, and whose covariance $K(A,B) = \Gamma(Z_A,Z_B) = \mu(A \cap B)$ is determined by μ. Such a process is an example of an uncorrelated random field.

2. Uncorrelated Random Fields

2.1. Definitions. Let $\mathbb{T} = [-\pi,\pi[$ and let T be the family of all Borel subsets of $\mathbb{T}$. Let (Ω,P) be a probability space. An uncorrelated random field Z on $\mathbb{T}$ (also called an uncorrelated random measure) is a map $A \to Z_A$ from T into $L^2(\Omega,P)$ such that

(i) if $A,B \in \mathcal{T}$ and $A \cap B = \emptyset$ then Z_A and Z_B are uncorrelated and $Z_{A \cup B} = Z_A + Z_B$;

(ii) if $A_n \in \mathcal{T}$ decreases toward the empty set as $n \to \infty$, then Z_{A_n} tends to 0 in L^2.

Note that if η is a (deterministic) complex valued, bounded measure on $\mathbb{T}$, then the map $A \to \eta(A)$ defines an uncorrelated "random" field which is in fact deterministic.

2.2. Centered Fields

Let Z be an uncorrelated random field. Write $\theta(A) = Z_A$ and $Z_A = Z'_A + \theta(A)$. Then Z'_A is *a centered uncorrelated random field* (i.e. $E(Z'_A) = 0$ for all A) *and* θ *is a bounded measure on* $\mathbb{T}$.

We shall only check this last point: the finite additivity of θ is a consequence of (i); on the other hand for $A \in \mathcal{T}$

$$|\theta(A)| = |E(Z_A)| \leq E|Z_A| \leq \|Z_A\|_2$$

and hence $\theta(A_n)$ tends to 0 whenever A_n decreases toward the empty set. The measure θ is bounded, since by (i),

$$\|Z_A\|_2^2 + \|Z_{A^c}\|_2^2 = \|Z_{\mathbb{T}}\|_2^2$$

whence $|\theta(A)| \leq \|Z_{\mathbb{T}}\|_2$.

From here on we shall limit ourselves to the use of centered uncorrelated random fields.

2.3. Theorem (and Definition). *Let* Z_A, $A \in \mathcal{T}$, *be a complex valued, centered, second-order process, indexed by the Borel subsets of* $\mathbb{T}$. *Then* Z *is an uncorrelated random field if and only if there is a bounded positive measure* μ *on* $(\mathbb{T},\mathcal{T})$ *such that the covariance of* Z *is given by*

$$\Gamma(Z_A,Z_B) = \mu(A \cap B). \tag{1}$$

The (deterministic) μ *is called the basis of the field* Z.

Proof. Assume Z to be an uncorrelated, field. Write

$\mu(A) = \sigma^2(Z_A)$. For $A,B \in \mathcal{T}$ and $A \cap B = \emptyset$, one has $\Gamma(Z_A,Z_B) = 0$ and hence

$$\mu(A \cup B) = \sigma^2(Z_{A\cup B}) = \sigma^2(Z_A + Z_B)$$
$$= \sigma^2(Z_A) + \sigma^2(Z_B) = \mu(A) + \mu(B).$$

If A_n decreases toward $\emptyset$, $\mu(A_n)$ tends to 0 since

$$\mu(A_n) = \|Z_{A_n}\|_2^2 .$$

Thus μ is a positive measure on $\mathbb{T}$, bounded since $\mu(\mathbb{T}) = \sigma^2(Z_{\mathbb{T}})$ is finite.

For arbitrary sets A and B, we may write $A = C \cup D$, $B = C \cup F$ with D, F, C disjoint and $C = A \cap B$ whence by (i) $Z_A = Z_C + Z_D$, $Z_B = Z_C + Z_F$, and Z_D, Z_F, Z_C pairwise orthogonal, which implies

$$\Gamma(Z_A,Z_B) = \Gamma(Z_C,Z_C) = \sigma^2(Z_{A\cap B}) = \mu(A \cap B).$$

Conversely let Z be a centered process verifying (1), with μ a bounded positive measure. To see that Z is additive we write

$$\sigma^2(Z_{A\cup B} - Z_A - Z_B) = \sigma^2(Z_{A\cup B}) + \sigma^2(Z_A) + \sigma^2(Z_B)$$
$$- 2\Gamma(Z_{A\cup B},Z_A) - 2\Gamma(Z_{A\cup B},Z_B) + 2\Gamma(Z_A,Z_B).$$

By (1) the right-hand side is equal to

$$\mu(A \cup B) - \mu(A) - \mu(B) + 2\mu(A \cap B) = \mu(A \cap B).$$

Consequently, if $A \cap B = \emptyset$ the r.v. $[Z_{A\cup B} - Z_A - Z_B]$ has zero variance, whence $Z_{A\cup B} = Z_A + Z_B$. This proves point (i) in Definition 2.1. We deduce the point (ii) from the fact that if A_n decreases toward $\emptyset$ the sequence

$$\|Z_{A_n}\|_2 = \sqrt{\mu(A_n)}$$

tends to zero, since μ is a measure.

2.4. Example. Uncorrelated Random Measure With Countable Support

Let A_k, $k \geqslant 1$, be a sequence of centered r.v., pairwise uncorrelated, belonging to $L^2(\Omega,P)$. Write $\sigma_k^2 = \sigma^2(A_k)$ and assume $\Sigma_{k\geqslant 1}\sigma_k^2$ to be finite. Let λ_k, $k \geqslant 1$, be a sequence of distinct points in $\mathbb{T}$. Then for all $B \subset \mathbb{T}$, *the series*

$$Z_B = \sum_{k\geqslant 1} A_k 1_B(\lambda_k)$$

converges in $L^2(\Omega,P)$ *and defines a centered uncorrelated random field* $Z = (Z_B)_{B\subset\mathbb{T}}$. *The basis* μ *of* Z *is given by*

$$\mu = \sum_{k\geqslant 1} \sigma_k^2 \delta_{\lambda_k} .$$

We shall say that Z *is an uncorrelated (centered) random measure with countable support* included in $\{\lambda_k, k \geqslant 1\}$, support which may of course be finite if the A_k are zero for $k \geqslant N$. We adopt *the notation*

$$Z = \sum_{k\geqslant 1} A_k \delta_{\lambda_k} .$$

This result is proved by checking, through a direct computation sketched in Section 1, that $\Gamma(Z_B,Z_C) = \mu(B \cap C)$ for $B,C \subset \mathbb{T}$. One then applies 2.3.

2.5. Example. Gaussian Random Fields

Every bounded positive measure μ *on* $(\mathbb{T},\mathcal{T})$ *is the basis of at least one centered uncorrelated random field* Z *on* $\mathbb{T}$, *and* Z *may be chosen to be a real valued gaussian field.*

Indeed, for $A_1 \dots A_r \in \mathcal{T}$, $x_1 \dots x_r \in \mathbb{R}$ one has

$$\sum_{i,j} x_i x_j \mu(A_i \cap A_j) = \int_{\mathbb{T}} \left[\sum_{i,j} x_i x_j 1_{A_i} 1_{A_j}\right] d\mu$$

$$= \int_{\mathbb{T}} \left|\sum_i x_i 1_{A_i}\right|^2 d\mu \geqslant 0$$

so that the function K: $\mathcal{T} \times \mathcal{T} \to \mathbb{R}$ given by $K(A,B) = \mu(A \cap B)$ is nonnegative definite (cf. Chapter 2, Section 4.4).

Consequently (cf. Chapter 2, Section 4.2) there is a *real valued centered gaussian process* $Z = (Z_A)_{A\in\mathcal{T}}$ indexed by $\mathcal{T}$, unique up to equivalence, with covariance $\Gamma(Z_A,Z_B) = K(A,B) = \mu(A \cap B)$. By Theorem 2.3, Z is an uncorrelated random field, with basis μ.

2.6. *Random Measures*

If Z is an uncorrelated random field on $\mathbb{T}$, Definition 2.1 implies that for every sequence A_n of *disjoint* Borel subsets of $\mathbb{T}$ one has, with $A = \cup_n A_n$

$$Z_A = \sum_n Z_{A_n} \quad \text{in} \quad L^2(\Omega,P)$$

where the series converges in L^2. *This does not imply*, in general, the possibility of choosing for each $A \in T$ a precise version $Y_A: \Omega \to \mathbb{C}$ of the P-equivalence class defined by $Z_A \in L^2$, in such a way that for P-almost all $\omega \in \Omega$ the map $A \to Y_A(\omega)$ is a (complex valued) measure on $\mathbb{T}$.

For instance, if β_t, $t \geqslant 0$, is the standard Brownian motion starting at 0 (cf. [15], Vol. 2) one can prove the existence of a centered uncorrelated random field Z on $\mathbb{T} = [-\pi,\pi[$ such that

$$Z_{]a,b]} = \beta_{\pi+b} - \beta_{\pi+a} \quad \text{for} \quad [a,b] \subset \mathbb{T}.$$

This **Brownian field** is a real valued gaussian field and its basis is Lebesgue's measure restricted to $\mathbb{T}$. But for almost all $\omega \in \Omega$, the map $A \to Z_A(\omega)$ is not a measure since the trajectories of β are not of bounded variation.

On the other hand if

$$Z = \sum_k A_k \delta_{\lambda_k}$$

is a random measure with countable support (cf. 3.4), then for P-almost all $\omega \in \Omega$, the formula

$$\nu_\omega = \sum_k A_k(\omega) \delta_{\lambda_k}$$

defines a (complex valued) measure, *if one assumes* $\Sigma_k \sigma_k$ to be finite, where $\sigma_k^2 = \sigma^2(A_k)$. Indeed, this condition implies the almost sure convergence of $\Sigma_k |A_k|$. Since $\omega \to \nu_\omega(B)$ is for each $B \subset \mathbb{T}$ a version of Z_B, we thus justify the terminology "random measure" adopted in 3.4. Note that this property is not necessarily true when one has only $\Sigma_k \sigma_k^2 < \infty$ but $\Sigma_k \sigma_k = \infty$.

We have prefered the terminology of random fields to the terminology of random measures, to avoid possibilities of mathematical confusion. From the intuitive point of view, the vocabulary of random measures remains nevertheless a useful guide.

3. Stochastic Integrals

Let Z be an uncorrelated field. We seek to define

$$Z_f = \int_{\mathbb{T}} f\, dZ \quad \text{for } f\colon \mathbb{T} \to \mathbb{C}.$$

When the map $A \to Z_A(\omega)$ is a measure ν_ω on $\mathbb{T}$ for P-almost all $\omega \in \Omega$, one can attempt to define (as in Section 1)

$$Z_f(\omega) = \int_{\mathbb{T}} f(\lambda) d\nu_\omega(\lambda).$$

But, as seen in 2.6, this approach cannot deal with the general case. We shall use instead a remark made in the case of measures with finite support: the map $f \to Z_f$ "must be isometric."

3.1. Theorem. *Let Z be a centered, uncorrelated random field on $\mathbb{T}$, with basis μ. Then there is a unique isometry $f \to Z_f$ from $L^2(\mathbb{T},\mu)$ into $L^2(\Omega,P)$ such that $Z_{1_A} = Z_A$ for every Borel subset A of $\mathbb{T}$.*

One has $E(Z_f) = 0$ for all $f \in L^2(\mathbb{T},\mu)$ and the image of $L^2(\mathbb{T},\mu)$ by Z is equal to the closed vector subspace H^Z of $L^2(\Omega,P)$ generated by the Z_A, $A \in \mathcal{T}$.

3.2. Definition. Let Z be a random field as in 3.1. For every function $f \in L^2(\mathbb{T},\mu)$ we define the stochastic integral of f with respect to the field Z to be the element Z_f of $L^2(\Omega,P)$ associated to f by the isometry 3.1. We shall write

$$Z_f = \int_{\mathbb{T}} f\, dZ = \int_{\mathbb{T}} f(\lambda) dZ(\lambda).$$

Theorem 3.1 may then be transcribed by the following relations, where $f,g \in L^2(\mathbb{T},\mu)$ and $u,v \in \mathbb{C}$,

$$\int_{\mathbb{T}} (uf + vg) dZ = u\int_{\mathbb{T}} f\, dZ + v\int_{\mathbb{T}} g\, dZ$$

$$E\left[\int_{\mathbb{T}} f\, dZ\right] = 0$$

$$E\left[\left(\int_{\mathbb{T}} f\, dZ\right)\overline{\left(\int_{\mathbb{T}} g\, dZ\right)}\right] = \int_{\mathbb{T}} f\bar{g}\, d\mu.$$

Proof of Theorem 3.1. Write $H_1 = L^2(\mathbb{T},\mu)$ and $H_2 = L^2(\Omega,P)$. For $A \in T$ define $v_A \in H_1$ and $w_A \in H_2$ by $v_A = 1_A$ and $w_A = Z_A$. Then for $A,B \in T$ one has

(2)
$$\langle v_A, v_B\rangle_{H_1} = \int_{\mathbb{T}} 1_A 1_B d\mu = \mu(A \cap B)$$
$$= \Gamma(Z_A, Z_B) = \langle w_A, w_B\rangle_{H_2}.$$

Let V and W respectively be the closed vector subspaces of H_1 and H_2 generated by the $(v_A)_{A\in T}$ and the $(w_A)_{A\in T}$. The step functions being dense in $L^2(\mathbb{T},\mu)$ (cf. [15], Chapter 3), one has $V = H_1$. On the other hand W is by definition the linear envelope H^Z of the field Z. By (2), Proposition 4.3.1 implies the existence of a unique isometry S from V *onto* W, hence from $L^2(\mathbb{T},\mu)$ onto H^Z, such that $Sv_A = w_A$, hence such that $S1_A = Z_A$. On the other hand, the process $Z = (Z_A)_{A\in T}$ being centered, one has $EY = 0$ for all $Y \in H^Z$ (cf. Chapter 4, Section 1). Whence Theorem 3.1.

3.3. The Case of Random Measures with Countable Support

On $\mathbb{T}$, consider a centered, uncorrelated random measure with countable support

$$Z = \sum_k A_k \delta_{\lambda_k},$$

with the notations of 2.4. The basis μ of Z being given by

$$\mu = \sum_k \sigma_k^2 \delta_{\lambda_k},$$

a function f: $\mathbb{T} \to \mathbb{C}$ is in $L^2(\mathbb{T},\mu)$ when

$$\sum_k \sigma_k^2 |f(\lambda_k)|^2 < \infty,$$

and one has then the converging series in $L^2(\Omega,P)$

$$Z_f = \int_{\mathbb{T}} f\, dZ = \sum_k A_k f(\lambda_k).$$

Indeed a very simple direct computation suffices to check that if one **defines** U_f as the sum of the series $\Sigma_k A_k f(\lambda_k)$, then

$$\Gamma(U_f, U_g) = \int_{\mathbb{T}} f\bar{g}\, d\mu ,$$

which forces $Z_f = U_f$ by 3.1.

We have associated an isometry from $L^2(\mathbb{T},\mu)$ into $L^2(\Omega,P)$ to every random field with basis μ. We now study the converse.

3.4. Theorem. *Let (Ω,P) be a probability space and let μ be a positive, bounded measure on $\mathbb{T}$. Let J be a linear map from $L^2(\mathbb{T},\mu)$ into $L^2(\Omega,P)$ such that $E[J(f)] = 0$ for all $f \in L^2(\mathbb{T},\mu)$. Then J is an isometry if and only if there is a centered uncorrelated random field on $\mathbb{T}$ such that*

$$J(f) = \int_{\mathbb{T}} f\, dZ.$$

Moreover J determines Z uniquely.

Proof. Start with an isometry J and define Z_A by $Z_A = J(1_A)$, which implies

$$\begin{aligned}\Gamma(Z_A,Z_B) &= \langle J(1_A),J(1_B)\rangle_{L^2(\Omega,P)}\\ &= \langle 1_A,1_B\rangle_{L^2(\mathbb{T},\mu)} = \mu(A \cap B)\end{aligned}$$

and by Theorem 2.3, Z is hence a centered, uncorrelated random field, with basis μ. Since $f \to Z_f = \int_{\mathbb{T}} f\, dZ$ is the unique isometry coinciding with Z_A for $f = 1_A$, we get $J(f) \equiv Z_f$ for all $f \in L^2(\mathbb{T},\mu)$.

Bibliographical Hints

A general study of Hilbertian and gaussian random fields may be found in *Neveu.*

Chapter VI
SPECTRAL REPRESENTATION OF STATIONARY PROCESSES

1. Processes with Finite Spectrum

Since we may identify the one dimensional torus with $\mathbb{T} = [-\pi,\pi[$ to each bounded measure ν on $\mathbb{T}$ we associate its Fourier transform

$$\hat{\nu}(n) = \int_{\mathbb{T}} e^{in\lambda} d\nu(\lambda), \qquad n \in \mathbb{Z}.$$

In particular if $Z = \Sigma_{k=1}^{r} A_k \delta_k$ is an uncorrelated random measure on $\mathbb{T}$, with finite support (cf. Chapter 5, Section 1), the Fourier transform of Z is obviously random and is given by

$$X_n = \hat{Z}(n) = \int_{\mathbb{T}} e^{in\lambda} dZ(\lambda) = \sum_k A_k e^{in\lambda_k}.$$

The process $X = \hat{Z}$ is second-order stationary (cf. Chapter 3, Section 5), and centered if Z is centered. Thus the centered second-order stationary processes having finite spectrum (cf. Chapter 3, Section 5) are the Fourier transforms of the centered uncorrelated random measures with finite spectrum.

Moreover, if we let $f_m(\lambda) = e^{im\lambda}$, we have $X_m = Z_{f_m}$ and the covariance of X becomes, μ being the basis of Z:

$$\gamma(m - n) = K(m,n) = \Gamma(X_m,X_n) = \Gamma(Z_{f_m},Z_{f_n})$$
$$= \int_{\mathbb{T}} f_m \bar{f}_n d\mu = \hat{\mu}(m - n)$$

using the isometry $f \to Z_f$.

Thus the function γ appears as the Fourier transform of the (deterministic) positive measure μ.

We are going to extend these results to all stationary random processes.

2. Spectral Measures

We begin by showing how to obtain all functions $\gamma: \mathbb{Z} \to \mathbb{C}$ such that $\gamma(m - n)$ is the covariance $K(m,n)$ of at least one second-order stationary process. Such functions are **nonnegative definite** (cf. Chapter 1, Section 4.4), i.e. satisfy for $r \in \mathbb{N}$, $x_1 \ldots x_r \in \mathbb{C}$, $n_1 \ldots n_r \in \mathbb{Z}$,

$$\sum_{k,\ell} x_k \bar{x}_\ell \gamma(n_k - n_\ell) \geq 0 \quad \text{and} \quad \gamma(n) = \overline{\gamma(-n)}.$$

2.1. Theorem (Bochner-Herglotz). *A function $\gamma: \mathbb{Z} \to \mathbb{C}$ is nonnegative definite if and only if there exists a positive bounded measure μ on $\mathbb{T} = [-\pi,\pi[$ such that*

$$\gamma(n) = \int_{\mathbb{T}} e^{in\lambda} d\mu(\lambda) \quad \textit{for all} \quad n \in \mathbb{Z}.$$

The function γ determines μ uniquely.

***Proof*.** If μ is a bounded positive measure on $\mathbb{T}$, it Fourier transform γ clearly satisfies $\gamma(n) = \overline{\gamma(-n)}$, and for $x_1 \ldots x_r \in \mathbb{C}$, we have

$$\sum_{k,\ell} x_k \bar{x}_\ell \gamma(n_k - n_\ell) = \int_{\mathbb{T}} \left| \sum_k x_k e^{in_k\lambda} \right|^2 d\mu(\lambda) \geq 0.$$

Thus γ is nonnegative definite.

We sketch the converse when

$$\sum_n |\gamma(n)| < \infty.$$

Write

$$h_N(\lambda) = \frac{1}{N} \sum_{1 \leq k,\ell \leq N} e^{ik\lambda} e^{-i\ell\lambda} \gamma(k - \ell)$$

so that $h_N \geq 0$ whenever γ is nonnegative definite. But

$$h_N(\lambda) = \sum_{-N \leqslant k \leqslant N} \left[1 - \frac{|k|}{N}\right] \gamma(k) e^{-ik\lambda}$$

tends to

$$h(\lambda) = \sum_k \gamma(k) e^{-ik\lambda} \qquad \text{as } N \to +\infty,$$

since the series are dominated in modulus by $\Sigma_k |\gamma(k)|$. The continuous function h is hence real valued, nonnegative, and the measure μ on $\mathbb{T}$ having the density $(1/2\pi)h$ with respect to Lebesgue measure is bounded and positive. An immediate computation gives $\hat{\mu}(n) = \gamma(n)$. For the general case, see [47], [37].

2.2. Definition. Let X be a second-order stationary random process, indexed by $\mathbb{Z}$. The **spectral measure** μ of X is the unique bounded positive measure μ on $\mathbb{T}$ such that the covariance of X is given by

$$\Gamma(X_m, X_n) = \int_{\mathbb{T}} e^{i(m-n)\lambda} d\mu(\lambda), \qquad m,n \in \mathbb{Z}.$$

Let us point out that if X is real valued, the measure μ is invariant by the symmetry s: $\mathbb{T} \to \mathbb{T}$ defined by $s(\lambda) = -\lambda$ modulo 2π.

2.3. Example: Processes with Countable Spectrum. Consider a centered stationary process X, with countable spectrum (cf. Chapter 3, Section 5), given by

$$X_n = \sum_k A_k e^{in\lambda_k},$$

where the λ_k are distinct points of $\mathbb{T}$ and the A_k are centered uncorrelated r.v., with variances σ_k^2 verifying $\Sigma_k \sigma_k^2 < \infty$. We have seen that the covariance $K(m,n)$ of X is given by

$$K(m,n) = \sum_k \sigma_k^2 e^{i(m-n)\lambda_k};$$

and hence $K(m,n) = \hat{\mu}(m - n)$ with

$$\mu = \sum_k \sigma_k^2 \delta_{\lambda_k}.$$

This explicitly give the spectral measure μ of X.

Here X is a linear combination, with uncorrelated random coefficients, of the deterministic periodic processes $n \to e^{in\lambda_k}$, and *the set of all frequencies* λ_k *of the periodic components of* X *is exactly the support of the spectral measure* μ *of* X. Of course the σ_k are supposed to be nonzero. If X is an optical or hertzian signal, the **spectrum** of X, that is the set of all the λ_k is a very concrete physical notion.

The mass σ_k^2 *given by the spectral measure* μ *to the frequency* λ_k *is the average energy* $E|A_k|^2$ *of the periodic component* $n \to A_k e^{in\lambda_k}$.

2.4. Spectral Densities

If the spectral measure μ of a process X has a density f with respect to Lebesgue measure on $\mathbb{T}$, we say that X has **spectral density** f. *When* X *is real valued, the function* f *is necessairly even*, i.e. satisfies $f(\lambda) = f(-\lambda)$ for almost all $\lambda \in \mathbb{T}$.

For X to have a (Lebesgue) square integrable density f, it is necessary and sufficient that the covariances $\gamma(n) = \hat{\mu}(n)$ verify $\Sigma_n|\gamma(n)|^2 < \infty$ and we then have

$$(1) \qquad f(\lambda) = \frac{1}{2\pi} \sum_n \gamma(n) e^{-in\lambda} \qquad \text{a.e. } \lambda \in \mathbb{T}$$

where the series converge in $L^2(\mathbb{T}, d\lambda)$. If we have the stronger condition $\Sigma_n|\gamma(n)| < \infty$, the Fourier series converges uniformly in λ and f is continuous (cf. 2.1). In fact the order of differentiability of f increases with the speed of convergence to zero (as $|n| \to \infty$) of the Fourier coefficients $\gamma(n)$ of $2\pi f$ (cf. Appendix).

Example: White Noise. Let W be a white noise with variance σ^2 (cf. Chapter 4, Section 2.2). By definition its covariances are $\gamma(0) = \sigma^2$ and $\gamma(n) = 0$ for $n \neq 0$. By (1), W has constant spectral density

$$f(\lambda) \equiv \frac{\sigma^2}{2\pi}.$$

All frequencies $\lambda \in \mathbb{T}$ contribute with "the same energy" to an eventual spectral decomposition of white noise, a fact which, by analogy with white light, justifies the "color" of W.

In practice, W is used to modelize imprevisible and impartial perturbations (random shocks in econometry, noise in signals transmission).

Example: Moving Averages. Let W be a white noise with variance σ^2 and call X the process defined by

$$X_n = \sum_k c_{n-k} W_k \qquad n \in \mathbb{Z},$$

where *the* $c_j \in \mathbb{C}$ *are zero for* $|j|$ *large enough.* Then X is centered second-order stationary since the covariance

$$\Gamma(X_m, X_n) = \sum_{k,\ell} c_{m-k} \bar{c}_{n-\ell} E(W_k \bar{W}_\ell) = \sigma^2 \sum_k c_{m-k} \bar{c}_{n-k}$$

may be written $\gamma(m - n)$ with

$$\gamma(n) = \sigma^2 \sum_k c_{n-k} \bar{c}_{-k} \, .$$

The sequence $\gamma(n)$ is zero for $|n|$ large enough and hence (1) proves that X has a spectral density f given by

$$f(\lambda) = \frac{1}{2\pi} \sum_n \gamma(n) e^{-in\lambda} = \frac{\sigma^2}{2\pi} \sum_n \sum_k c_{n-k} \bar{c}_{-k} e^{-in\lambda}.$$

The change of indices $k = -\ell$, $n = m - \ell$ implies

$$f(\lambda) = \frac{\sigma^2}{2\pi} \sum_m \sum_\ell c_m \bar{c}_\ell e^{-i(m-\ell)\lambda} = \frac{\sigma^2}{2\pi} \left| \sum_m c_m e^{-im\lambda} \right|^2.$$

The process X is called a **moving average.** We shall see (Chapter 7) that the preceding computation remains valid provided $\sum_m |c_m|^2 < \infty$.

3. Spectral Decomposition

3.1. Proposition. *Let* Z *be an uncorrelated centered random field on* $\mathbb{T}$, *having the basis* μ. *Then the Fourier transform* X *of* Z *defined by*

$$X_n = \int e^{in\lambda} dZ(\lambda), \qquad n \in \mathbb{Z},$$

is a centered, second-order stationary process, with spectral measure μ.

Proof. Let $f_m(\lambda) = e^{im\lambda}$ so that (cf. Chapter 5) $X_m = Z_{f_m}$. Since $f \to Z_f$ is isometric, we have

$$\Gamma(X_m,X_n) = \langle Z_{f_m}, Z_{f_n}\rangle_{L^2(\Omega,P)} = \langle f_m,f_n\rangle_{L^2(\mathbb{T},\mu)}$$
$$= \int_{\mathbb{T}} e^{i(m-n)\lambda} d\mu(\lambda)$$

whence the result.

Let us prove that **every** stationary process is the Fourier transform of a random field.

3.2. Theorem. *Let X be a second-order stationary process with spectral measure μ. Then there exists on $\mathbb{T}$ a unique centered uncorrelated random field Z such that $X = \hat{Z}$, i.e. such that*

$$X_n = \int_{\mathbb{T}} e^{in\lambda} dZ(\lambda), \qquad n \in \mathbb{Z}.$$

Moreover the field Z then has the basis μ.

We shall call Z **the spectral field** of X.

Proof. Let $H_1 = L^2(\mathbb{T},\mu)$, $H_2 = L^2(\Omega,P)$, and $f_m(\lambda) = e^{im\lambda}$. In H_1 and H_2 respectively consider the vector sequences $(f_n)_{n\in\mathbf{Z}}$ and $(X_n)_{n\in\mathbf{Z}}$. The definition of spectral measures implies

$$(2) \qquad \langle f_m,f_n\rangle_{H_1} = \int_{\mathbb{T}} e^{i(m-n)\lambda} d\mu(\lambda) = \langle X_m,X_n\rangle_{H_2} \quad \text{for } m,n \in \mathbb{Z}$$

Let F_1, F_2 be the closed vector subspaces of H_1, H_2 respectively generated by the (f_m) and the (X_n). By (2) and Proposition 3.1 of Chapter 4, there exists a unique isometry J of F_1 **onto** F_2 such that $J(f_n) = X_n$ for all $n \in \mathbb{Z}$. But by definition (cf. Chapter 4, Section 1) F_2 is the linear envelope H^X of X, and $F_1 = H_1$ since the finite linear combinations of complex exponentials are dense in $L^2(\mathbb{T},\mu)$ (cf. Appendix). Moreover, X being centered we have (cf. Chapter 4, Section 1) $EY = 0$ for all $Y \in H^X$.

Theorem 3.4 (Chapter 5) prove that the isometry J from $L^2(\mathbb{T},\mu)$ into $L^2(\Omega,P)$ is of the form

$$J(f) = \int_{\mathbb{T}} f\, dZ, \quad f \in L^2(\mathbb{T},\mu),$$

where Z is a centered uncorrelated random field on $\mathbb{T}$, with basis μ. In particular,

$$X_n = J(f_n) = \int_{\mathbb{T}} e^{in\lambda} dZ(\lambda).$$

If Z' is another uncorrelated random field such that $X = \hat{Z}'$, Proposition 3.1 shows that Z' also has the basis μ. The isometries $f \to Z_f$ and $f \to Z'_f$ from H_1 into H_2 coincide for $f = f_n$, and hence coincide on the closed vector subspace $F_1 = H_1$ generated by the f_n, so that $Z \equiv Z'$.

3.3. Proposition. *Let X be a centered second-order stationary process. Let Z be its spectral field, and μ its spectral measure. Then the linear envelopes H^X and H^Z coincide, and the isometry $f \to \int_{\mathbb{T}} f\, dZ$ maps $L^2(\mathbb{T},\mu)$ onto $H^X = H^Z$.*

Proof (Notations 3.2). We have just seen that the isometry J maps $L^2(\mathbb{T},\mu)$ onto H^X. But the isometry $f \to Z_f = \int_{\mathbb{T}} f\, dZ$ maps $L^2(\mathbb{T},\mu)$ onto H^Z, by 3.1, Chapter 5. Since these isometries coincide by construction, the result is proved.

3.4. The Case of Real Valued Processes

It is useful to point out that *even if the process X is real valued its spectral field is not real valued in general.*

Indeed, when we identify the one-dimensional torus with $\mathbb{T} = [-\pi,\pi[$, the natural symmetry of the torus (passage from $e^{i\lambda}$ to $e^{-i\lambda}$) becomes the map s: $\mathbb{T} \to \mathbb{T}$ defined by $s(\lambda) \equiv -\lambda$ (mod 2π). If θ is a bounded complex measure on $\mathbb{T}$, we define the measure θ^* by the formula

$$(3) \qquad \int_{\mathbb{T}} f\, d\theta^* = \int_{\mathbb{T}} f \circ s\, d\bar{\theta}$$

for all bounded Borel functions f: $\mathbb{T} \to \mathbb{C}$, where $\bar{\theta}$ is the conjugate measure defined by $\bar{\theta}(A) = \overline{\theta(A)}$, $A \subset \mathbb{T}$.

If Z is a centered uncorrelated random field on $\mathbb{T}$, with basis μ, we have for A, B Borel subsets of $\mathbb{T}$

$$\Gamma(\bar{Z}_A,\bar{Z}_B) = \overline{\Gamma(Z_A,Z_B)} = \mu(A \cap B)$$

so that $A \to \bar{Z}_A$ defines (Chapter 5, Theorem 2.3) another centered uncorrelated random field, *having the same basis μ as Z.* We shall hence denote $\bar{Z}$ the **conjugate field of** Z which obviously satisfies

$$(4) \qquad \overline{\left[\int_{\mathbb{T}} f\, dZ\right]} = \int_{\mathbb{T}} \bar{f}\, d\bar{Z} \qquad f \in L^2(\mathbb{T},\mu).$$

The random process Y obtained by Fourier transform of $\bar{Z}$ is not *the conjugate of $X = \hat{Z}$ but is linked to it by time inversion*, i.e.

$$Y_n = \int e^{in\lambda} d\bar{Z}(\lambda) \quad \text{and} \quad X_n = \int e^{in\lambda} dZ(\lambda)$$

verify $Y_n = \bar{X}_{-n}$.

On the other hand, if $s(\mu)$ is the image of μ by the symmetry s, the map $f \to f \circ s$ from $L^2(\mathbb{T},s(\mu))$ onto $L^2(\mathbb{T},\mu)$ is an isometry, and hence the map $f \to \int_{\mathbb{T}} f \circ s\, dZ$, being the composition of two isometries, is an isometry of $L^2(\mathbb{T},s(\mu))$ into $L^2(\Omega,P)$, with centered values. Theorem 3.4, Chapter 5 implies the existence of a unique centered random field that we shall denote $s(Z)$, having the basis $s(\mu)$, such that

$$(5) \qquad \int_{\mathbb{T}} f\, d[s(Z)] = \int_{\mathbb{T}} f \circ s\, dZ \qquad f \in L^2(\mathbb{T},s(\mu)).$$

We shall say that $s(Z)$ is the **symmetric field** of Z, and by (5) it is clear that *the process V with spectral field $s(Z)$ is linked to $X = \hat{Z}$ by time inversion*, i.e. $V_n = X_{-n}$.

By (4) and (5) we may define the **adjoint field**, Z^* *of Z by*

$Z^* = \overline{s(Z)} = s(\bar{Z})$ *which has the basis $s(\mu)$ and satisfies the analogue of* (3)

$$\int_{\mathbb{T}} f\, dZ^* = \int f \circ s\, d\bar{Z} \qquad f \in L^2(\mathbb{T},s(\mu)).$$

If the spectral field of X is Z, then Z^* is the spectral field of $\bar{X}$.

The uniqueness of spectral fields shows that the relation $X = \bar{X}$ is equivalent to $Z = Z^*$; *a centered stationary process X is hence real valued if and only if its spectral field Z is self-adjoint*, i.e. *verifies $Z = Z^*$.*

But random processes X with real valued spectral fields $Z = \bar{Z}$ are exactly those which verify $X_{-n} \equiv X_n$ and hence do not correspond to usual random stationary phenomena.

3.5. Example: Processes with Countable Spectrum.

Let

$$X = \sum_k A_k e^{in\lambda_k}$$

be a stationary process with countable spectrum (cf. 2.3). It is easily checked that *the spectral field Z of X is the uncorrelated random measure with countable support* (cf. Chapter 5, Section 2.4)

$$Z = \sum_k A_k \delta_{\lambda_k} .$$

In particular X is real if and only if $Z = Z^*$, that is, if

$$Z = A_{-1}\delta_{-\pi} + A_0\delta_0 + \sum_{j\geq 1} A_j\delta_{\lambda_j} + \sum_{j\geq 1} \bar{A}_j\delta_{-\lambda_j}$$

where the λ_j, $j \geq 1$ are distinct points of $]0,\pi[$, A_{-1} and A_0 are centered real valued r.v. and the A_j, $j \geq 1$ are centered complex r.v. such that

$$E(A_k\bar{A}_\ell) = 0 \qquad \text{for } k \neq \ell \qquad k,\ell \geq -1$$

$$E(A_j^2) = 0 \qquad \text{for } j \geq 1.$$

Letting

$$A_k = \frac{1}{2}(U_k - iV_k)$$

one then has

$$X_n = U_0 + (-1)^n U_{-1} + \sum_{k\geq 1} (U_k \cos n\lambda_k + V_k \sin n\lambda_k).$$

Bibliographical Notes

Bochner's theorem and basic notions on Fourier series may be found in *Rudin.* Let us mention also *Neveu* (1). The main result on Fourier series are recalled in the Appendix.

Chapter VII
LINEAR FILTERS

1. Often Used Linear Filters

1.1. Finite Moving Averages

Let X be a centered second-order stationary process. Let a_k, $k \in \mathbb{Z}$ be a sequence of complex numbers, equal to zero for $|k|$ large enough. The process $Y = AX$ defined by

$$Y_n = \sum_k a_k X_{n-k}$$

is then centered and second-order stationary, for if $\gamma(m-n) = \Gamma(X_m, X_n)$, the covariance of Y becomes

$$\Gamma(Y_m, Y_n) = \sum_{k,\ell} a_k \bar{a}_\ell E(X_{m-k} \bar{X}_{n-\ell}) = \sum_{k,\ell} a_k \bar{a}_\ell \gamma(m-n-k+\ell)$$

which is clearly a function of $(m-n)$.

The operation $X \to Y = AX$, called **finite moving average** is used often in practice (smoothing of irregular observations, elimination of trends,...) and is an important particular case of linear filter. Let us give two examples.

1.2. Elimination of Polynomial Trends

Let Y be a random process of the form $Y_n = f(n) + X_n$ where X is centered stationary and f is a deterministic polynomial of degree r. Write $\Delta f(n) = f(n) - f(n-1)$, $\Delta Y_n = Y_n - Y_{n-1}$.

Clearly, we have $\Delta^{r+1}f \equiv 0$ and

$$\Delta^{r+1}Y = \sum_{k=0}^{r+1} a_k Y_{n-k}$$

where the a_k are fixed integers. Moreover $\Delta^{r+1}Y = \Delta^{r+1}X$ and is hence (by 1.1) centered stationary. The moving average Δ^{r+1} has eliminated the polynomial trend.

1.3. Elimination of Periodic Trends

Let Y be a process of type $Y_n = f(n) + X_n$ where X is centered stationary and $f(n)$ is a deterministic linear combination of $\sin((2\pi/r)nj)$ and $\cos(2\pi/r)nj)$, $j = 0,1, \ldots (r - 1)$. Thus f is periodic with period r; in numerous concrete cases (econometrics, meteorology) one has $r = 12$ for monthly observations, and f is called the **seasonal component** of Y.

As is well known, every root z of $z^r - 1 = 0$ satisfies, provided $z \neq 1$, $z^n + z^{n-1} + \ldots + z^{n-r+1} = 0$. Since $f(n)$ is a linear combination (with constant coefficients) of the z^n where z runs through the set of roots of $z^r - 1 = 0$, we must have

$$Af(n) = \frac{1}{r} \sum_{j=0}^{r-1} f(n-j) \equiv a$$

where a is a constant. Consequently the process

$$AY = \frac{1}{r} \sum_{j=0}^{r-1} Y_{n-j}$$

may be written $AY = a + AX$, and hence is second-order stationary. The moving average A has eliminated the periodic trend; one says that it has deseasonalized Y. We could have considered $BY = Y_n - Y_{n-r}$ which would have had an analogous effect.

1.4. Forecasting

Let X be a stationary process. To forecast the value of X_{n+1} when the X_k, $k \leqslant n$ are known, we shall use (cf. Chapter 9) a linear combination of the X_{n-k} of type $\hat{X}_{n+1} = \Sigma_{k\geqslant 0} a_k X_{n-k}$, that is a slightly more general linear filter than in 1.1.

1.5. The Response Function of a Finite Moving Average

As in 1.1 let us consider a centered stationary process X and its image $Y = AX$ by a **finite** moving average. Call Z^X, Z^Y the spectral fields of X, Y. We then have

$$Y_n = \sum_k a_k X_{n-k} = \int_{\mathbb{T}} \left[\sum_k a_k e^{i(n-k)\lambda}\right] dZ^X(\lambda)$$

$$= \int_{\mathbb{T}} e^{in\lambda} h(\lambda) dZ^X(\lambda)$$

with

$$h(\lambda) = \sum_k a_k e^{-ik\lambda}, \qquad \lambda \in \mathbb{T}.$$

Since

$$Y_n = \int_{\mathbb{T}} e^{in\lambda} dZ^Y(\lambda)$$

it is tempting to write **formally** "$dZ^Y(\lambda) = h(\lambda) dZ^X(\lambda)$."

The function h: $\mathbb{T} \to \mathbb{C}$ is called the response function of the moving average A. We shall give a meaning to the formalism $dZ^Y = h\, dZ^X$ to present a mathematical definition of general linear filters.

2. Multiplication of a Random Field by a Function

2.1. Densities

Let V and Z be (centered) random fields on $\mathbb{T}$, with respective bases ν and μ. By analogy with the terminology of measure theory, we shall say that *V has a density $h \in L^2(\mathbb{T},\mu)$ with respect to Z* if

$$\int_{\mathbb{T}} f\, dV = \int_{\mathbb{T}} fh\, dZ \quad \text{for all} \quad f \in L^2(\mathbb{T},\nu)$$

which we shall denote by $dV = h\, dZ$. The isometry properties of stochastic integrals (cf. Chapter 5) imply then, for $f \in L^2(\mathbb{T},\nu)$

$$\int_{\mathbb{T}} |f|^2 d\nu = \mathrm{var}\left[\int_{\mathbb{T}} f\, dV\right] = \mathrm{var}\left[\int_{\mathbb{T}} fh\, dZ\right]$$

$$= \int_{\mathbb{T}} |f|^2 |h|^2 d\mu.$$

Consequently the measure ν must have the density $|h|^2$ with respect to μ, which may be denoted by $d\nu = |h|^2 d\mu$.

2.2. Proposition. *Let Z be a centered uncorrelated random field on* $\mathbb{T}$, *with basis* μ. *Then for every function h in* $L^2(\mathbb{T},\mu)$ *there exists a unique centered uncorrelated random field V on* $\mathbb{T}$ *such that* $dV = h\,dZ$. *The basis* ν *of V then has density* $|h|^2$ *with respect to* μ.

Proof. Given h, we consider a priori the bounded positive measure ν on $\mathbb{T}$ such that $d\nu = |h|^2 d\mu$. The map $f \to hf$ is an isometry of $L^2(\mathbb{T},\nu)$ into $L^2(\mathbb{T},\mu)$ since

$$\int (hf)(\overline{hg})d\mu = \int f\bar{g}|h|^2 d\mu = \int f\bar{g}\, d\nu.$$

Since $\varphi \to Z_\varphi = \int_{\mathbb{T}} \varphi\, dZ$ is an isometry of $L^2(\mathbb{T},\mu)$ into $L^2(\Omega,P)$, we conclude that $f \to Z_{hf}$ is an isometry of $L^2(\mathbb{T},\nu)$ into $L^2(\Omega,P)$, with $E(Z_{hf}) = 0$. By Chapter 5, Theorem 3.4 there is then a unique centered uncorrelated random field V with basis ν such that $V_f = Z_{hf}$ for all $f \in L^2(\mathbb{T},\nu)$.

3. Response Functions and Linear Filters

3.1. Theorem and Definition. *Let X be a second-order stationary process, with spectral measure* μ^X *and spectral field* Z^X. *Let* $h \in L^2(\mathbb{T},\mu)$. *Then the centered process Y defined by*

$$(1) \qquad Y_n = \int_{\mathbb{T}} e^{in\lambda} h(\lambda) dZ^X(\lambda)$$

is second-order stationary. Its spectral measure μ^Y *admits with respect to* μ^X *the density* $|h|$. *Its spectral field* Z^Y *has density h with respect to* Z^X.

We say that *Y is the* **image** *of X but the linear filter with response function.*

Proof. Let V be the random field defined by $dV = h\,dZ^X$. The basis ν of V is then given by $d\nu = |h|^2 d\mu^X$. By (1) and 2.1, we have $Y = \hat{V}$, and hence $V = Z^Y$, $\nu = \mu^Y$.

3.2. Proposition. *Let Y be the image of X by the filter with* *response function h, where X, h are as in* 3.1. *The linear envelope of Y and X are then related by* $H^Y \subset H^X$. *One has* $H^Y = H^X$ *if and only if* $h > 0$ *μ-almost everywhere; in this case X is the image of Y by the filter with response function* $1/h$.

Proof. Let J: $L^2(\mathbb{T},\mu^Y) \to L^2(\mathbb{T},\mu^X)$ be the isometry $J(f) = hf$. The isometries Z^Y and Z^X of $L^2(\mathbb{T},\mu^Y)$ and $L^2(\mathbb{T},\mu^X)$ into $L^2(\Omega,P)$ verify $Z^Y = Z^X \circ J$ by 2.1. But H^Y and H^X are the respective images of $L^2(\mathbb{T},\mu^Y)$ and $L^2(\mathbb{T},\mu^X)$ by Z^Y and Z^X. This implies $H^Y \subset H^X$, and, isometries being necessarily one-to-one, the relation $H^Y = H^X$ is equivalent to the fact that J is onto.

Clearly J can only be onto if $h > 0$ μ-almost everywhere. It is then invertible and

$$J^{-1}g = \frac{1}{h}g \quad \text{for} \quad g \in L^2(\mathbb{T},\nu).$$

The relation $Z^X = Z^Y \circ J^{-1}$ may then be written (by 2.1)

$$dZ^X = \frac{1}{h}\, dZ^Y,$$

which concludes the proof.

3.3. Example: Band-Pass Filter. This is the filter with response function $h = 1_B$ where B is a Borel subset of $\mathbb{T}$. The image Y of X by such a filter has its spectral measure carried by the "band" B, and hence only involves the frequencies λ belonging to the "band" B. The undesirable frequencies λ, i.e., the λ which do not belong to B have been eliminated by this filter.

3.4. Example: Infinite Moving Averages. Let X be a centered stationary process, with spectral measure μ and spectral field Z. Let a_k, $k \in \mathbb{Z}$ be a sequence of complex numbers. Write $f_m = e^{im\lambda}$, so that $X_m = Z_{f_m}$. The map $f \to Z_f$ being isometric, convergence in $L^2(\Omega,P)$ of the series

$$Y_n = \sum_k a_k X_{n-k} = \sum_k a_k Z_{f_{n-k}}$$

is equivalent to convergence in $L^2(\mathbb{T},\mu)$ for the series $g_n = \Sigma_k a_k f_{n-k}$, and we then have $Y_n = Z_{g_n}$. But $g_n = f_n g_0$ so that *the series* $Y_n = \Sigma_k a_k X_{n-k}$ *converges in* $L^2(\Omega,P)$ *if and only if*

$$h(\lambda) = g_0(\lambda) = \sum_k a_k e^{-ik\lambda}$$

converges in $L^2(\mathbb{T},\mu)$. In this case we have

$$Y_n = Z_{g_n} = Z_{hf_n} = \int_{\mathbb{T}} e^{in\lambda} h(\lambda) dZ(\lambda)$$

and *Y is the image of X by the filter with response function h.*

This filter is often called an **infinite moving average**. Let us point out that if $\Sigma_k |a_k| < \infty$ the convergence of

$$h(\lambda) = \sum_k a_k e^{-ik\lambda}$$

in $L^2(\mathbb{T},\mu)$ is true for all bounded positive measures μ.

If X has a bounded spectral density, the weaker condition $\Sigma_k |a_k|^2 < \infty$ which classically implies (cf. Appendix) the convergence of

$$h(\lambda) = \sum_k a_k e^{-ik\lambda}$$

in $L^2(\mathbb{T}$, Lebesgue measure), clearly forces the convergence in $L^2(\mathbb{T},\mu)$.

3.5. Product Filters

Let $(\Omega,\mathcal{B})$ be a space of trials. Let $S(\Omega)$ be the set of all centered second-order stationary processes defined on $(\Omega,\mathcal{B})$, with values in $\mathbb{C}$, and indexed by the time $\mathbb{Z}$.

Let $\Omega_0 = \mathbb{C}^{\mathbb{Z}}$, endowed with its usual product σ-algebra. Then for every $X' \in S(\Omega)$ there is an $X \in S(\Omega_0)$ which is equivalent to X'. This is a consequence of Kolmogorov's theorem 1.3.4.

From now on we fix $S = S(\Omega_0)$. To every measurable function $h: \mathbb{T} \to \mathbb{C}$ we associate an operator $F_h: \mathcal{D}_h \to S$, with domain $\mathcal{D}_h \subset S$, which is the linear filter with response function h.

The domain of definition $\mathcal{D}_h$ of F_h is the set of all $X \in S$ with spectral measures μ^X verifying

$$\int_{\mathbb{T}} |h|^2 d\mu^X < \infty.$$

For instance *if h is bounded, one has* $\mathcal{D}_h = S$. Let us point out that if $X \in \mathcal{D}_h \cap \mathcal{D}_g$, the relation $F_h X = F_g X$ is equivalent to $h = g$, μ^X-a.e.

The natural domain of definition for the product $F_h F_g$ is

$$\mathcal{D}[h,g] = \{X \in S \mid X \in \mathcal{D}_g \text{ and } F_g X \in \mathcal{D}_h\}$$

i.e. the set of $X \in S$ such that $\int_{\mathbb{T}} |g|^2 d\mu^X$ and $\int_{\mathbb{T}} |h|^2 |g|^2 d\mu^X$ are finite. We note that $\mathcal{D}[h,g]$ and $\mathcal{D}[g,h]$ do not generally coincide as shown by the example $g(\lambda) = \lambda - \lambda_0$ and $h(\lambda) = 1/\lambda - \lambda_0$ for $\lambda \neq \lambda_0$ and $h(\lambda_0)$ arbitrary.

Proposition. *Let h,g: $\mathbb{T} \to \mathbb{C}$ be measurable functions. The natural domain of definition $\mathcal{D}[h,g]$ for $F_h F_g$ is included in $\mathcal{D}_{hg}$ and one has $F_h F_g X = F_{hg} X$ for all $X \in \mathcal{D}[h,g]$.*

Proof. If $X \in \mathcal{D}[h,g]$ then $\int_{\mathbb{T}} |g|^2 d\mu^X$ and $\int_{\mathbb{T}} |h|^2 |g|^2 d\mu^X$ are finite. Thus $|hg|$ is in $L^2(\mathbb{T}, \mu^X)$ and $X \in \mathcal{D}_{hg}$. Write $U = F_g X$, $V = F_h U$ and $Y = F_{hg} X$. The definition of filters shows that the spectral fields verify

$$dZ^U = g\, dZ^X, \quad dZ^V = h\, dZ^U, \quad dZ^Y = hg\, dZ^X.$$

Definition 2.1 then easily implies $Z^V = Z$ which by Fourier transform gives $V = Y$ and proves the proposition.

3.6. Inverse Filters

We adopt the following convention; for every function $h: \mathbb{T} \to \mathbb{C}$, we define $[1/h]$: $\mathbb{T} \to \mathbb{C}$ by

$$\left[\frac{1}{h}\right](\lambda) = \frac{1}{h(\lambda)}$$

when $h(\lambda) \neq 0$, and $[1/h](\lambda) = 0$ when $h(\lambda) = 0$. We shall call $F_{[1/h]}$ the **inverse filter** of F_h. Its domain of definition $\mathcal{D}_{[1/h]}$ is the set of all $X \in S$ such that

$$\int_{\mathbb{T}} 1_{\{h \neq 0\}} \frac{1}{|h|^2} d\mu^X$$

is finite. In particular *if h has no zero on* $\mathbb{T}$ we have $F_{[1/h]} = F_{1/h}$.

The natural domain of definition of $F_{[1/h]}F_h$ *is clearly equal to* $\mathcal{D}_h$ and the preceding proposition shows that for all $X \in \mathcal{D}_h$, one has $F_{[1/h]}F_hX = F_\varphi X$ where $\varphi = 1_{\{h \neq 0\}}$.

The relation $X = F_{[1/h]}F_hX$ *is hence only valid when* $h \neq 0$ μ^X-a.e. *and* $X \in \mathcal{D}_h$. For instance for h and $1/h$ bounded, the inverse filter of F_h is $F_{1/h}$ and one has $X = F_{1/h}F_hX$ for all $X \in \mathcal{S}$.

4. Applications to Linear Representations

4.1. Proposition. *Let* X *be a centered stationary process. Then the following properties are equivalent:*

(i) *X has a spectral density f*

(ii) *there is a sequence* $c_k \in \mathbb{C}$ *verifying* $\Sigma_k|c_k|^2 < \infty$ *and such that X has the spectral density*

$$f(\lambda) = \left|\sum_{k\in\mathbf{Z}} c_k e^{-ik\lambda}\right|^2, \qquad \lambda \in \mathbb{T}.$$

(iii) *there is a sequence* $c_k \in \mathbb{C}$ *verifying* $\Sigma_k|c_k|^2 < \infty$ *and a white noise W such that*

$$X_n = \sum_{k\in\mathbf{Z}} c_k W_{n-k}\,.$$

Proof. Call η the Lebesgue measure on $\mathbb{T}$, μ^X the spectral measure of X, Z^X its spectral field. Start from (i). Let $g\colon \mathbb{T} \to \mathbb{C}$ be an arbitrary Borel function such that

$$(2) \qquad |g|^2 = f.$$

Since $\int_{\mathbb{T}} |g|^2 d\lambda$ is finite, the Fourier coefficients

$$c_k = \frac{1}{2\pi}\int_{\mathbb{T}} e^{ik\lambda} g(\lambda)d\lambda$$

satisfy $\Sigma_k|c_k|^2 < \infty$ and

$$(3) \qquad g = \sum_{k\in\mathbf{Z}} c_k f_{-k}$$

where $f_k(\lambda) = e^{ik\lambda}$, and the series (3) converges in $L^2(\mathbb{T},\eta)$. This proves (ii).

Start from (ii), and consider first the case $f > 0$ almost everywhere. The measures μ^X and η are then equivalent by hypothesis and $|g| = f$ is η-a.e. nonzero. The function $[1/g]$ defined in 3.6 is in $L^2(\mathbb{T},\mu^X)$ and the filter $F_{[1/g]}$ is hence defined on X. The spectral measure of the process $W = F_{[1/g]}X$ is given by

$$d\mu^W = |[1/g]|^2 d\mu^X = 1_{\{g\neq 0\}} \frac{f}{|g|^2} d\eta = d\eta.$$

Thus W is a white noise with variance 2π. Since $g \neq 0$ η-a.e. we have $X = F_g F_{[1/g]} X$ and hence $X = F_g W$. The spectral density of W is bounded and $\Sigma_k |c_k|^2 < \infty$; by 3.4 the expansion (3) of g shows that X may be written

$$(4) \qquad X_n = \sum_{k\in\mathbb{Z}} c_k W_{n-k} \qquad n \in \mathbb{Z}$$

where the series converges in $L^2(\Omega,P)$, whence (iii).

The case $\eta[\{g = 0\}] > 0$ is more involved. The theorem remains true but the linear envelope of X is then strictly contained in the linear envelope of W. We shall not study this case here (cf. [15]).

Finally (iii) implies $X = F_g W$ with g given by (3) whence

$$d\mu^X = |g|^2 d\mu^W = \frac{\sigma^2}{2\pi} |g|^2 d\eta$$

which proves (i), and 4.1 is proved.

The linear representation (4) *of X is far from being unique. there are as "as many" such representations as choices of g: $\mathbb{T} \to \mathbb{C}$ verifying $|g|^2 = f$.*

4.2. Proposition. *Let X be a centered stationary process. The following properties are equivalent:*

(i) *X is regular*

(ii) *X has a spectral density f of the form*

$$f(\lambda) = \left| \sum_{k\geq 0} c_k e^{-ik\lambda} \right|^2, \qquad \lambda \in \mathbb{T},$$

where the $c_k \in \mathbb{C}$ verify $\Sigma_{k\geq 0}|c_k|^2 < \infty$.

(iii) *there is a sequence $c_k \in \mathbb{C}$ satisfying $\Sigma_{k\geq 0}|c_k|^2 < \infty$ and a white noise w such that*

$$X_n = \sum_{k\geq 0} c_k W_{n-k} \,.$$

Proof. If X is regular, its innovation W verifies (iii) by Chapter 4, Theorem 3.4. Conversely if W is a white noise linked to X by (iii) we have $X_m \in H_n^W$ for all $m \leq n$, hence $H_n^X \subset H_n^W$ and $H_{-\infty}^X \subset H_{-\infty}^W$. This last subspace is reduced to $\{0\}$ since a white noise is always regular. Thus X is regular and (i) is equivalent to (iii). The equivalence of (iii) and (ii) is a consequence of 4.1.

4.3. Theorem (Kolmogorov). *Let X be a centered stationary process. Then X is regular if and only if X has a spectral density f such that $\log f$ is Lebesgue integrable. The one step forecasting error σ^2 satisfies*

$$\log \sigma^2 = \frac{1}{2\pi} \int_{\mathbb{T}} \log 2\pi f(\lambda) d\lambda.$$

Proof. For the general case see ([15, Vol. 2]). Further on (cf. Chapter 9) we shall study the important particular case where $\log f$ has Fourier coefficients converging to 0 fast enough. the idea of the proof is to discover $g = \Sigma_{k \geq 0} c_k e^{-ik\lambda}$ such that $|g|^2 = f$. In the "nice" cases one writes

$$\log f(\lambda) = \sum_{k \in \mathbb{Z}} u_k e^{-ik\lambda},$$

and one defines

$$h(\lambda) = \frac{1}{2} u_0 + \sum_{k \geq 1} u_k e^{-ik\lambda}$$

so that $h + \bar{h} = \log f$ and $g = e^h$ verifies $|g|^2 = f$. When it is legitimate to compute formally the Fourier series of e^h by replacement of h by its Fourier series in $\Sigma_{r \geq 0} (1/r!) h^r$, it is clear that the Fourier coefficients of g are zero for $k < 0$.

5. *Characterization of Linear Filters as Operators*

If X is a centered stationary process, and F_h a filter defined on X, we may associate to F_h a linear operator A_h defined on the finite linear combinations of the X_n by

$$A_h\left(\sum_k c_k X_k\right) = \sum_k c_k A_h X_k = \sum_k c_k Y_k$$

where $Y = F_h X$. Let us characterize these linear operators.

5.1. *Definitions*. Let H be a complex Hilbert space and $\mathcal{D}$ a dense vector subspace of H. A linear operator A: $\mathcal{D} \to H$ is said to be **closed** if for every sequence $v_n \in \mathcal{D}$, the simultaneous convergence of v_n to $v \in H$ and Av_n to $w \in H$ implies $v \in \mathcal{D}$ and $Av = w$.

The adjoint $(A^*, \mathcal{D}^*)$ of $(A, \mathcal{D})$ is defined by $\mathcal{D}^* = \{u \in H \mid$ there is a $w \in H$ satisfying $\langle w,v\rangle = \langle u,Av\rangle$ for all $v \in \mathcal{D}\}$ and $\langle A^*u,v\rangle = \langle u,Av\rangle$ for $u \in \mathcal{D}^*$, $v \in \mathcal{D}$. Finally $(A, \mathcal{D})$ is **normal** whenever $AA^* = A^*A$.

5.2. *Theorem*. *Let X be a centered stationary process with spectral measure μ and let T: $H^X \to H^X$ be the backward shift of X. Let $\mathcal{D}$ be a vector subspace of H^X and A: $\mathcal{D} \to H$ a linear operator such that*

(i) $X_n \in \mathcal{D}$ *for all* $n \in \mathbb{Z}$; $T\mathcal{D} \subset \mathcal{D}$; $AT = TA$;
(ii) A *is closed and normal.*

Then the process Y defined by $Y_n = AX_n$ is centered and stationary, and there exists a unique function $h \in L^2(\mathbb{T},\mu)$ such that

(iii) *Y is the image of X by the filter with response function h.*

All linear filters may be described as above since to each $h \in L^2(\mathbb{T},\mu)$ one may associate $(A, \mathcal{D})$ verifying (i), (ii), and (iii).

Proof (sketched). Let Z be the spectral field of X, which defines an isometry of $\hat{H} = L^2(\mathbb{T},\mu)$ onto $H = H^X$. Write $f_n(\lambda) = e^{in\lambda}$ to get $X_n = Z(f_n)$. The backward shift T being isometric, the operator $\hat{T}$: $\hat{H} \to \hat{H}$ defined by $\hat{T} = Z^{-1}TZ$ is an isometry verifying $\hat{T}f = f_1 f$.

Start with $h \in \hat{H}$. To prove the second assertion of the theorem, we must exhibit $(A, \mathcal{D})$ verifying (i), (ii), and (iii). It is obviously enough to find $\hat{A} = Z^{-1}AZ$ and $\hat{\mathcal{D}} = Z^{-1}(\mathcal{D})$. Simple verifications show that $\hat{A}f = hf$ and $\hat{\mathcal{D}} = \{f \in \hat{H} \mid hf \in \hat{H}\}$ have the desired properties since $\hat{A}$ is closed, normal, commutes trivially with $\hat{T}$, and $\hat{A}f_n = hf_n$ implies $AX_n = Z(hf_n)$.

Conversely, given $(A, \mathcal{D})$ verifying (i) and (ii), we must exhibit $h \in \hat{H}$ such that $AX_n = Z(hf_n)$. By the isometry Z, this amounts to the characterization of all closed, normal $(\hat{A}, \hat{\mathcal{D}})$ which commute with $\hat{T}$ and such that $\hat{\mathcal{D}}$ includes all the f_n. Setting $h = \hat{A}f_0$ and comparing $\hat{A}$ with the operator $\hat{B}$ of multiplication by h, we conclude that $\hat{A} = \hat{B}$ after noticing that $f_n = \hat{T}^n f_0$.

Bibliographical Notes

One important theoretical result which has not been proved here is Kolmogorov's theorem on regularity. The proof refered to in [15] is essentially inspired from *Hoffman*. A very complete probabilistic approach can be found in *Ibraguimov* and *Rozanov*. From a different point of view, the notion of filter is widely used in engineering (signal theory, electronics, etc.). We shall only mention a few books in English on the subject: *Eyckhoff*, *Franklin-Powell*, *Astrom-Wittenmark*, where the numerical properties of filters are studied.

Chapter VIII
ARMA PROCESSES AND PROCESSES WITH RATIONAL SPECTRUM

1. ARMA Processes

1.1. Definitions. Let X be a stationary centered process, with values in $\mathbb{R}$ or $\mathbb{C}$, defined on (Ω,P). We shall say that X **is an ARMA (p,q) process** if there exists a white noise W defined on (Ω,P) and real or complex numbers $a_0 ... a_p$, $b_0 ... b_q$ such that

$$(1) \qquad \sum_{k=0}^{p} a_k X_{n-k} = \sum_{\ell=0}^{q} b_\ell W_{n-\ell}, \qquad n \in \mathbb{Z}.$$

Let us describe more precisely two important particular cases.

One says that X is **autogressive of order p**, or more briefly **an AR(p) process**, if it satisfies (1) with $q = 0$, that is if

$$(2) \qquad \sum_{k=0}^{p} a_k X_{n-k} = W_n, \qquad n \in \mathbb{Z}.$$

One says that X is **a moving average of order** q, or more briefly an MA(q) **process** if it verifies (1) with $p = 0$, that is if

$$(3) \qquad X_n = \sum_{\ell=0}^{q} b_\ell W_{n-\ell}, \qquad n \in \mathbb{Z}.$$

This terminology has been popularized by Box-Jenkins [10].

1.2. Example. In econometrics, one often considers dynamic models of the following type; call $x_1 ... x_r$ the numerical characteristics of interest in a given arbitrary economic sector (prices, levels of production, incomes, investments, etc.), assumed to be yearly observable for instance. To the year n is then associated a vector $X(n)$ with coordinates $x_1(n)...x_r(n)$. One assumes that $X(n)$ verifies a linear recursion of the type

$$X(n) = A_0X(n) + A_1X(n-1) + ... + A_jX(n-j)$$

where $A_0...A_j$ are matrices, but that this "ideal" relation is perturbed by random effects $W(n)$ with zero means, uncorrelated for different years. Thus the $X(n)$ become random vectors verifying

$$X(n) = \sum_{k=0}^{j} A_kX(n-k) + W(n). \tag{4}$$

One may show, with a few technical restrictions on the A_k, that if $W(n)$ is a (vector) white noise, then there is a stationary process $X(n)$ verifying (4) such that for each $i = 1...r$, $x_i(n)$ is an ARMA process.

1.3. ARMA Relation and Backward Shifts

Let X be an ARMA process; then there is a white noise W linked to X by (1). Let us point out that *in general the linear envelopes H^X and H^W do not coincide and that a fortiori the two backward shifts T^X: $H^X \to H^X$ and T^W: $H^W \to H^W$ are distinct; even if $H^X = H^W$, nothing guarantees a priori that T^X and T^W coincide.* One may of course rewrite (1) in condensed form

$$P(T^X)X_n = Q(T^W)W_n$$

where P and Q are the polynomials

$$P(z) = \sum_{k=0}^{p} a_kz^k \qquad Q(z) = \sum_{\ell=0}^{q} b_\ell z^\ell \tag{5}$$

but $P(T^X)$ and $Q(T^W)$ remain operators which are defined on distinct spaces. A sizeable amount of the ARMA literature peacefully ignores these difficulties by assuming implicitly or not that $H^X = H^W$ and $T^X = T^W$.

1.4. ARMA Relation and Linear Filters

Let $\mathbb{C}[z]$ be the ring of polynomials with complex coefficients. To every $P \in \mathbb{C}[z]$ we asociate *the bounded continuous function* h_P *defined on* $\mathbb{T} = [-\pi,\pi[$ *by*

(6) $$h_P(\lambda) = P(e^{-i\lambda}) \qquad \lambda \in \mathbb{T}$$

and *we denote* A_P *the filter* F_{h_P} *with response function* h_P, which is defined on every stationary process. By Chapter 7, A_P is a finite moving average which verifies

$$A_P X_n = \sum_{k=0}^{p} a_k X_{n-k} \quad \text{for} \quad P(z) = \sum_{k=0}^{p} a_k z^k.$$

For $P_1,P_2 \in \mathbb{C}[z]$ and $P = P_1 P_2$ one has $h_P = h_{P_1} h_{P_2}$ and hence (Chapter 7) $A_P = A_{P_1} A_{P_2}$.

With these conventions and P, Q as in (5), the typical ARMA relation (1) between X and white noise W becomes

(6) $$A_P X = A_Q W.$$

For every polynomial R one has then $A_{PR} X = A_{QR} W$.

2. Regular and Singular Parts of an ARMA Process

If $P \in \mathbb{C}[z]$ has no root of modulus 1, the function $1/h_P$ is bounded continuous and by Chapter 7, Section 3.6, the filter F_{1/h_P} is a **true inverse** of $F_{h_P} = A_P$, so that A_P^{-1} exists and is a filter defined on every stationary process. This suggests the following factorization.

2.1. Factorization of P

For every polynomial $P \in \mathbb{C}[z]$ there exists a factorization $P = P_r P_s$ unique up to a multiplicative constant, such that P_r has no root of modulus 1 and P_s has all its roots of modulus 1. We shall call P_r and P_s respectively **the regular and singular factors of** P. **We denote** $\text{sp}(P)$ **the set of** $\lambda \in \mathbb{T}$ **such that**

$P(e^{-i\lambda}) = 0$ so that $\mathrm{sp}(P) = \mathrm{sp}(P_s)$ and $\mathrm{sp}(P_r) = \emptyset$.

2.2. Proposition. *Let* $P,Q \in \mathbb{C}[z]$ *and* $P = P_r P_s$ *the factorization of P in regular and singular parts. Assume there exists a white noise W, not identical to zero, and a stationary centered process X linked by* $A_P X = A_Q W$. *Then* P_s *must divide Q and the spectral measure* μ^X *of X may be written* $\mu^X = \mu_r + \mu_s$ *where*

(i) *the support of* μ_s *is finite and included in* $\mathrm{sp}(P)$

(ii) μ_r *has density* $\dfrac{d\mu_r(\lambda)}{d\lambda} = \left|\dfrac{Q}{P}(e^{-i\lambda})\right|^2, \ \lambda \in \mathbb{T}$,

with respect to Lebesgue measure.

Proof. Since $A_P X = A_Q W$, the spectral measures of $A_P X$ and $A_Q W$ coincide with a measure ν such that

$$d\nu = |h_P|^2 d\mu^X \tag{7}$$

$$d\nu = |h_Q|^2 d\mu^W = \frac{\sigma^2}{2\pi}|h_Q|^2 d\eta \tag{8}$$

where η is Lebesgue measure and σ^2 is the variance of W.

Let $\mu^X = \mu_r + \mu_s$ be the Radon-Nikodym decomposition of μ^X with respect to η (cf. [15, Vol. I, p. 270]) with absolutely continuous part μ_r and singular part μ_s. By (7) the Radon-Nikodym decomposition $\nu = \nu_r + \nu_s$ of ν with respect to η verifies

$$d\nu_r = |h_P|^2 d\mu_r \quad \text{and} \quad d\nu_s = |h_P|^2 d\mu_s \tag{9}$$

and hence by (8) we have $\nu = \nu_r$ and $\nu_s = 0$, whence

$$d\nu = d\nu_r = \frac{\sigma^2}{2\pi}|h_Q|^2 d\eta \ . \tag{10}$$

By (9) the relation $\nu_s = 0$ implies that μ_s is carried by $\{\lambda \in \mathbb{T} \mid h_P(\lambda) = 0\}$ that is by $\mathrm{sp}(P)$.

By definition, $d\mu_r = f\, d\eta$, and hence (9) implies

$$f|h_P|^2 = \frac{\sigma^2}{2\pi}|h_Q|^2,$$

η-a.e., and since h_P is nonzero η-a.e., we obtain

$$f = \frac{\sigma^2}{2\pi} \left|\frac{h_Q}{h_P}\right|^2,$$

η-a.e. But μ_r being bounded, $\int_{\mathbb{T}} f(\lambda)d\lambda$ is finite, and it is easily checked that

$$\int_{\mathbb{T}} \left|\frac{Q}{P}(e^{-i\lambda})\right|^2 d\lambda$$

is finite if and only if every factor $(z - \tau)^m$ of $P(z)$ with $|\tau| = 1$ is also a factor of $Q(z)$. Consequently the polynomial P_s, which is a product of such factors, must divide Q.

2.3. Theorems. *Let $P,Q \in \mathbb{C}[z]$, and let $P = P_rP_s$ be the factorization of P in regular and singular parts. Assume that there exists a centered stationary process X and a white noise W with nonzero variance, linked by $A_PX = A_QW$. One then has the factorization $Q = Q_1P_s$ and the unique decomposition $X = X^r + X^s$ where X^r and X^s are centered stationary processes verifying*

(i) *X^r and X^s are uncorrelated,* i.e., $\Gamma(X^r_m, X^s_n) = 0$ *for all m,n.*
(ii) *X^s has finite spectrum included in* sp(P)
(iii) *X^r has a spectral density and verifies $A_PX^r = A_QW$.*

Conversely, start with two arbitrary centered stationary processes X^r, X^s and a white noise W satisfying (i), (ii) *and* (iii). *Then $X = X^r + X^s$ is centered, stationary, and verifies $A_PX = A_QW$. Moreover, we then have*

(j) $A_{P_s}X^s = A_PX^s = 0$

(jj) $A_{P_r}X^r = A_{Q_1}W$

(jjj) *the spectral density f^r of X^r is given by*

$$f^r(\lambda) = \frac{\sigma^2}{2\pi} \left|\frac{Q}{P}(e^{-i\lambda})\right|^2$$

where $\sigma^2 =$ variance (W).

Proof. Let μ^X be the spectral measure of X and Z^X its spectral field. By 2.2, one has $\mu^X = \mu_r + \mu_s$ and the support G_s of μ_s is a (finite) subset of sp(P). Let $G_r = \mathbb{T} - G_s$ and call F_r, F_s the band-pass filters with response functions 1_{G_r} and 1_{G_s}. Define $X^r = F_rX$, $X^s = F_sX$ whence $X = X^r + X^s$ since

$$1_{G_r} + 1_{G_s} \equiv 1.$$

By construction,

$$d\mu^{X^s} = 1_{G_s} d\mu^X \quad \text{and} \quad d\mu^{X^r} = 1_{G_r} d\mu^X,$$

whence $\mu^{X^s} = \mu_s$ and then $\mu^{X^r} = \mu^X - \mu_s = \mu_r$. In particular X^s has finite spectrum and X^r has a spectral density.

Let $f_m(\lambda) = e^{im\lambda}$. By construction one has

$$X^r_m = \int_{\mathbb{T}} f_m 1_{G_r} dZ^X \quad \text{and} \quad X^s_n = \int_{\mathbb{T}} f_n 1_{G_s} dZ^X$$

whence, by isometry, the covariance

$$\Gamma(X^r_m, X^s_n) = \langle f_m 1_{G_r}, f_n 1_{G_s} \rangle_{L^2(\mathbb{T},\mu)} = 0$$

for all $m,n \in \mathbb{Z}$, since G_r and G_s are disjoint.

The function $h_{P_s}(\lambda) = P_s(e^{-i\lambda})$ is zero on sp(P). For every process U with spectral measure μ^U carried by sp(P), we must then have $A_{P_s} U = 0$; indeed the spectral measure χ of $A_{P_s} U$ verifies

$$d\chi = |h_{P_s}|^2 d\mu^U$$

and is hence equal to zero. In particular, we see that $A_{P_s} X^s = 0$, whence, a fortiori, $A_P X^s = 0$.

The relations $A_P X = A_Q W$ and $X = X^r + X^s$ then imply $A_P X^r = A_Q W$. Thus X^r and X^s satisfy (i), (ii), and (iii).

Conversely, consider two centered stationary processes U^r, U^s and a white noise W satisfying (i), (ii), and (iii). The centered process $X = U^r + U^s$ is stationary thanks to (i). By (ii) the spectral measure of U^s is carried by sp(P) whence, as seen above, $A_{P_s} U^s = 0$ and $A_P U^s = 0$. Since (iii) implies $A_P U^r = A_Q W$, we conclude that $A_P X = A_Q W$.

The relation $A_P U^r = A_Q W$ implies

(11) $$A_{P_s} Y = A_{P_s} V \quad \text{with} \quad Y = A_{P_r} U^r \quad \text{and} \quad V = A_{Q_1} W.$$

Since Y and V have spectral densities, the finite set of all $\lambda \in \mathbb{T}$ such that $h_{P_s}(\lambda) = 0$ is μ^Y- and μ^V-negligible.

Let F_φ, with

$$\varphi = \left[\frac{1}{h_{P_s}}\right]$$

be the inverse filter of F_s (cf. Chapter 7, Section 3.6), which must then satisfy $F_\varphi F_s Y = Y$ and $F_\varphi F_s V = V$. By (11) this forces $Y = V$ since $F_s Y = F_s V$. Whence

$$A_{P_r} U^r = A_{Q_1} W.$$

This proves (j) and (jj).

Since P_r has no roots of modulus 1, $A_{P_r}^{-1}$ exists and

$$U^r = A_{P_r}^{-1} A_{Q_1} W$$

is completely determined by P, Q, W; in particular $U^r = X^r$, whence $U^s = X^s$, and the uniqueness of the decomposition $X = X^r + X^s$. Finally assertion (jjj) is then a consequence of 2.2.

2.4. Wold's Decomposition

We have seen that processes with finite spectrum are singular (cf. Chapter 4) and we shall prove further on that the ARMA processes having a spectral density are regular. **The decomposition $X = X^r + X^s$ is hence Wold's decomposition** (cf. Chapter 4) **of X in regular and singular parts.**

The study of ARMA processes now clearly splits up into two parts. The first is very simple, it's the study of **processes with finite spectrum.** The other part is the study of ARMA processes having a spectral density, which are called **processes with rational spectrum** in view of Theorem 4.2 below.

2.5. The Equation $A_P X = A_Q W$ with X and W Given

By Theorem 2.3 *if X is an arbitrary process with finite spectrum, the only* ARMA *relations verified by X are of the type $A_P X = 0$ where P is an arbitrary polynomial such that $P(e^{-i\lambda}) = 0$ whenever λ belongs to the (finite) support of μ^X.*

Consider now *an* ARMA *process X having a spectral density, and a white noise W linked to X by an* ARMA *relation. Then* *there exists an irreducible fraction Q/P, where P has no zero of modulus* 1, *such that $A_P X = A_Q W$; moreover, the relation $A_{\tilde{P}} X = A_{\tilde{Q}} W$ then holds if and only if $\tilde{Q}/\tilde{P} = Q/P$.*

Indeed, start from the assumption $A_{\tilde{P}} X = A_{\tilde{Q}} W$. Note that W is nonzero, if not X would have finite spectrum. Theorem 2.3 implies $\tilde{P} = \tilde{P}_r \tilde{P}_s$ and $\tilde{Q} = \tilde{Q}_1 \tilde{P}_s$, as well as $A_{\tilde{P}_r} X = A_{\tilde{Q}_1} W$ since $X = X^r$. Let D be the greatest common divisor of $\tilde{P}_r$ and $\tilde{Q}_1$, and write $\tilde{P}_r = DP$, $\tilde{Q}_1 = DQ$. The roots of $\tilde{P}_r$ and hence those of D and P have moduli different from 1, so that A_D^{-1} exists. Multiplying by A_D^{-1} we get $A_P X = A_Q W$ which proves the announced statement since $\tilde{Q}/\tilde{P} = Q/P$.

3. Construction of ARMA Processes

3.1. The Equation $A_P X = A_Q W$ with Given P, Q, W

Given two polynomials $P,Q \in \mathbb{C}[z]$ and white noise W with nonzero variance, we want to find all the centered *stationary* solutions X of

$$(12) \qquad A_P X = A_Q W.$$

Let $P = P_r P_s$ be the factorization 2.1 of P in regular and singular parts. By Theorem 2.3, *for* (12) *to admit a stationary solution X, we must have $Q = P_s Q_1$, and then the general stationary solution of* (12) *is of the form*

$$X = A_{P_r}^{-1} A_{Q_1} W + Y$$

where Y is an arbitrary process with finite spectrum, uncorrelated with W, and such that support(μ^Y) *is included in* sp(P). The process $X^s = Y$ and $X^r = A_{P_r}^{-1} A_{Q_1} W$ are then uncorrelated.

In particular, *if P has no root of modulus* 1, *equation* (12) *has a unique stationary solution $X = A_P^{-1} A_Q W$.* We shall see that it is then possible to express the filter $A_P^{-1} A_Q$ as an infinite moving average.

Let us point out that *when X has a spectral density and verifies $A_P X = A_Q W$ with P, Q arbitrary, one always has* $H^X = H^W$. Indeed, with the notations of Chapter 7, Section 3.6, if we let $\varphi = [1/h_P]$, $\psi = [1/h_Q]$, we have, finite sets being μ^X-

and μ^W-negligible, $X = F_\varphi A_Q W$ and $W = F_\psi A_P X$. These relations imply (by Chapter 7, Proposition 3.2) $H^X \subset H^W$ and $H^W \subset H^X$ whence $H^X = H^W$.

3.2. Processes with Rational Spectrum and Infinite Moving Averages

Consider $P,Q \in \mathbb{C}[z]$ *with* P *having no root of modulus* 1. *Then* Q/P *admits the Laurent expansion*

$$\frac{Q}{P}(z) = \sum_{k \in \mathbb{Z}} c_k z^k \tag{13}$$

converging for $\alpha < |z| < 1/\alpha$ *where* $\alpha < 1$ *is fixed*, $z \in \mathbb{C}$, *and the coefficients* c_k *tend to* 0 *at exponential speed as* $|k| \to +\infty$.

Indeed, Q/P is the sum of a polynomial and a finite family of "simple fractions" of type $F(z) = c/(z - \tau)^r$ where τ is a root of P with multiplicity $\geq r$. One writes

$$(z - \tau)^{-r} = (-1)^r \tau^{-r}(1 - z/\tau)^{-r} \quad \text{if } |\tau| > 1$$

$$(z - \tau)^{-r} = z^{-r}(1 - \tau/z)^{-r} \quad \text{if } |\tau| < 1$$

and one uses the classical "binomial" expansion, converging for $|u| < 1$,

$$(1 - u)^{-r} = \sum_{k \geq 0} \frac{(r + k - 1)!}{r!k!} u^k.$$

This gives the two explicit expansions

$$\text{when } |\tau| > 1, \quad \frac{1}{(z-\tau)^r} = \sum_{k \geq 0} v_k z^k \quad \text{converging for } |z| < |\tau|$$

$$\text{when } |\tau| < 1, \quad \frac{1}{(z-\tau)^r} = \sum_{k \leq -r} w_k z^k \quad \text{converging for } |z| > |\tau|$$

and the bounds $|v_k| \leq \beta^{|k|}$ as well as $|w_k| \leq \beta^{|k|}$ with $0 < \beta < 1$. Whence the announced expansion of Q/P.

A Crucial Remark. *If* P *has all its roots of modulus* > 1, *then the coefficients* c_k *of the Laurent expansion of* Q/P *are zero for* $k < 0$.

Let us return to the case where P is only assumed to have

no root of modulus 1. *The unique stationary solution in X of $A_P X = A_Q W$ may then be written as an explicit infinite moving average*

$$X_n = \sum_{k \in \mathbf{Z}} c_k W_{n-k}$$

where the c_k are given by (13). Indeed the response function of the filter $A_P^{-1} A_Q$ is given by

$$h_P^{-1}(\lambda) h_Q(\lambda) = \frac{Q}{P}(e^{-i\lambda}) = \sum_{k \in \mathbf{Z}} c_k e^{-ik\lambda}$$

with $\Sigma_k |c_k| < \infty$. Whence the result by Chapter 7, Section 3.4.

Clearly the same argument applies to P/Q if Q has no root of modulus 1 and yields for W the expansion

$$W_n = \sum_{k \in \mathbf{Z}} d_k X_{n-k},$$

where the coefficients d_k are those of the Laurent expansion of P/Q.

The case where P and Q have all their roots of modulus > 1 is particularly interesting (cf. Section 5) since these linear representations are then both causal, i.e. $c_k = d_k = 0$ for $k < 0$, and X may be considered a process of the type AR($+\infty$) as well as of the type MA($+\infty$).

4. Processes with Rational Spectrum

4.1. Definition. A centered stationary process X is said to have **rational spectrum** if X has a spectral density of the form $f(\lambda) = F(e^{-i\lambda})$, $\lambda \in \mathbb{T}$, where F is a rational fraction with coefficients in $\mathbb{C}$.

Obviously F cannot be arbitrary since f must be real valued, positive, and integrable.

4.2. Theorem. *A centered stationary process X is an* ARMA *process having a spectral density if and only if X has rational spectrum.*

4.3. Corollary. *A centered stationary process X is an* ARMA *process if and only if its spectral measure μ^X may be written $\mu^X = \mu_r + \mu_s$, where μ_s has finite support and μ_r has a density of the form $d\mu_r/d\lambda = F(e^{-i\lambda})$, where F is a rational fraction with complex coefficients.*

Proofs. Start with an ARMA process X. The computation of μ^X given in 2.2 proves the direct assertions in 4.3 and 4.2. To establish the converse assertions, we only need to prove, thanks to Theorem 2.3, that if X has rational spectrum, then X verifies an ARMA relation. This will be a consequence of Theorem 4.6 and 4.7 below. Let us begin by examining the form of the density of X.

4.4. Theorem (Fejer-Riesz). *Let F be a complex rational fraction. The function $f(\lambda) = F(e^{-i\lambda})$ is real valued, positive, and Lebesgue integrable on $\mathbb{T}$, if and only if there exists an irreducible fraction Q/P, where P has no root of modulus 1, such that*

$$F(z) = \left|\frac{Q}{P}(z)\right|^2 \quad \text{for } |z| = 1, \quad z \in \mathbb{C}.$$

If F has real coefficients then one may choose Q, P with real coefficients.

Proof*. Factorizing F we have

$$F(z) = \alpha z^{r_0} \prod_{j \in J} (z - z_j)^{r_j}$$

with $\alpha \in \mathbb{C}$, r_0 and the r_j in $\mathbb{Z}$, and distinct nonzero $z_j \in \mathbb{C}$. The fraction $F(z)$ being real for $|z| = 1$, we have in this case, since $\bar{z} = 1/z$

$$F(z) = \bar{\alpha}\, \bar{z}^{r_0} \prod_{j \in J} (\bar{z} - \bar{z}_j)^{r_j} = \beta z^{-r_0 - \Sigma_{j \in J} r_j} \prod_{j \in J} (z - 1/\bar{z}_j)^{r_j}$$

with $\beta \in \mathbb{C}$. Two rational fractions in z which coincide for $|z| = 1$ must be identical, whence $\alpha = \beta$, and the fact that the map $(z_j, r_j) \to (1/\bar{z}_j, r_j)$ is a bijection of G onto G, where $G = \{(z_j, r_j)\}_{j \in J}$. This implies

$$G = \{(z_k, r_k)\}_{k \in K} \cup \{(1/\bar{z}_j, r_k)\}_{k \in K} \cup \{(z_\ell, r_\ell)\}_{\ell \in L}$$

where $|z_k| \neq 1$ for $k \in K$ and $|z_\ell| = 1$ for $\ell \in L$.

For $|z| = 1$ and z close to a given z_ℓ with $\ell \in L$, write $z = e^{-i\lambda}$, $z_\ell = e^{-i\lambda_\ell}$; then $F(e^{-i\lambda})$ is, for λ close to λ_ℓ, equivalent to $\gamma(\lambda - \lambda_\ell)^{r_\ell}$ with $\gamma \in \mathbb{C}$. For $F(e^{-i\lambda})$ to remain real and positive r_ℓ must be **even,** and the integrability of $F(e^{-i\lambda})$ in the neighborhood of λ_ℓ forces $r_\ell \geq 0$. We shall set $r_\ell = 2s_\ell$ where the s_ℓ are nonnegative integers, whence

$$(z - z_\ell)^{r_\ell} = (z - z_\ell)^{s_\ell}(z - 1/\bar{z}_\ell)^{s_\ell} \quad \text{since } |z_\ell| = 1.$$

Finally we see that F verifies

$$F(z) = \alpha z^{r_0} \prod_{\ell \in L} (z-z_\ell)^{s_\ell}(z - 1/\bar{z}_\ell)^{s_\ell} \prod_{k \in K} (z-z_k)^{r_k}(z-1/\bar{z}_k)^{r_k}.$$

The elementary identity, valid for $u,z \in \mathbb{C}$, $u \neq 0$, $|z| = 1$

$$(14) \qquad (z - u)(z - 1/\bar{u}) \equiv -\frac{1}{\bar{u}} z|z - u|^2$$

then shows that for $|z| = 1$ we have $F(z) = Mz^m|H(z)|^2$ with $M \in \mathbb{C}$, $m \in \mathbb{Z}$,

$$H(z) = \prod_{\ell \in L} (z - z_\ell)^{s_\ell} \prod_{k \in K} (z - z_k)^{r_k}.$$

Is is then obvious that F remains real valued and positive for $|z| = 1$ if and only if $m = 0$ and $M \in \mathbb{R}^+$, which proves the theorem, the fraction $H(z)$ having no pole of modulus 1, since $s_\ell \geq 0$.

The Fejer-Riesz representation is far from unique; the lack of uniqueness comes from the following identity, a consequence of (14)

$$(15) \qquad |z - u|^2 = |u|^2|z - 1/\bar{u}|^2 \quad \text{for } u,z \in \mathbb{C},\ u \neq 0,\ |z| = 1.$$

This point can be made more precisely.

4.5. Lemma. *The three following families of complex rational fractions,* $U_\alpha(z) \equiv \alpha$ with $|\alpha| = 1$, $V_m(z) = z^m$ *with* $m \in \mathbb{Z}$,

$$F_u(z) = |u| \frac{z - 1/\bar{u}}{z - u} \qquad \textit{with} \quad u \neq 0,\ u \in \mathbb{C}$$

verify $|U_\alpha(z)| \equiv |V_m(z)| \equiv |F_u(z)| \equiv 1$ *for* $|z| \equiv 1$. *Two irreducible rational fractions* Q/P *and* Q_1/P_1 *verify*

$$\left|\frac{Q}{P}(z)\right| \equiv \left|\frac{Q_1}{P_1}(z)\right|$$

for $|z| \equiv 1$ *if and only if* $Q/P = \Phi \times Q_1/P_1$ *where* Φ *is an arbitrary finite product of fractions of the type* U_α, V_m, *or* F_u.

Proof. It is sufficient to identify the fractions Φ such that $|\Phi(z)| \equiv 1$ for $|z| \equiv 1$. Factorizing Φ, one uses the remark

$$\bar{z} - \bar{u} = \frac{1}{z} - \bar{u} \quad \text{for} \quad |z| = 1$$

and the fact that rational fractions in z must be identical if they coincide for $|z| = 1$.

4.6. Theorem. *Let f be a real valued, positive Lebesgue integrable function on $\mathbb{T}$ such that $f(\lambda) = F(e^{-i\lambda})$, $\lambda \in \mathbb{T}$ where F is a complex rational fraction. Then f admits an infinity of irreducible Fejer-Riesz representations*

$$f(\lambda) = \left|\frac{Q}{P}(e^{-i\lambda})\right|^2 .$$

Starting from any such representations, all the others are obtained by 4.5. On the other hand, f admits a canonical Fejer-Riesz representation determined up to a multiplicative constant by the following extra requirement:

(16) *the irreducible fraction Q/P has all its poles of modulus > 1 and all its roots of modulus $\geqslant 1$.*

Proof. By 4.4, 4.5 we only need to exhibit the canonical representation of f. Start with an arbitrary irreducible representation

$$f(\lambda) = \left|\frac{Q}{P}(e^{-i\lambda})\right|^2,$$

where P has no root of modulus 1. By 4.5, replacing in P or Q any factor $(z - u)$ by $|u|(z - 1/\bar{u})$ does not change

$$\left|\frac{Q}{P}(z)\right| \quad \text{for} \quad |z| = 1.$$

This allows the replacement of poles and roots of modulus < 1 by poles and roots of modulus > 1.

When f is the spectral density of a process X, the following theorem associates to every Fejer-Riesz representation of f a white noise W and an ARMA relation linking X and W.

4.7. Theorem. *Let X be a process with rational spectrum; call f the spectral density of X. Let*

$$f(\lambda) = \left|\frac{Q}{P}(e^{-i\lambda})\right|^2$$

be an arbitrary irreducible Fejer-Riesz representation of f. Then there exists a white noise W with variance 2π *linked to X by the* ARMA *relation* $A_P X = A_Q W$.

Proof. Let $\psi = [1/h_Q]$ with the notations of Chapter 7, Section 3.6. One has $f = |h_Q/h_P|^2$ and $Y = A_P X$ has spectral density

$$f^Y = |h_P|^2 f = |h_Q|^2.$$

The filter F_ψ with response function ψ is hence defined on $A_P X$ and the process $W = F_\psi A_P X$ has spectral density $f^W \equiv 1$ since the zeros of h_Q form a finite set. Thus W is a white noise with variance 2π, and $W = F_\rho X$ with $\rho = \psi h_P$. We then have $A_Q W = A_Q F_\rho X$ and $A_Q F_\rho$ has a response function $h_Q \rho = h_Q \psi h_P$ which coincides with h_P almost everywhere. Whence $A_Q F_\rho X = A_P X$ and $A_P X = A_Q W$.

Thus *every process with rational spectrum verifies an infinity of irreducible* ARMA *relations*, which correspond bijectively to the irreducible Fischer-Riesz representations of its spectral density f. *The white noise associated to these* ARMA *relations are generally all distinct.*

We shall call **canonical ARMA relation** satisfied by X the ARMA relation associated to the canonical Fejer-Riesz representation of f. More precisely, one can, *in a unique fashion*, write

$$f(\lambda) = \frac{\sigma^2}{2\pi}\left|\frac{Q}{P}(e^{-i\lambda})\right|^2$$

where P, Q have no common factor, P has all its roots of modulus > 1, Q has all its roots of modulus $\geqslant 1$, and $P(0) = Q(0) = 1$.

The normalizing constants have been adjusted to give variance σ^2 to the white noise W linked to X by $A_P X = A_Q W$.

The correspondence between the canonical triples (σ^2, P, Q) as above and the covariance structure of X is obviously one-to-one. If p,q are the degrees of P, Q we shall say that X is of **minimal type** (p,q). We shall see that the white noise associated to the canonical ARMA relation is the innovation of X.

5. Innovation for Processes with Rational Spectrum

5.1. Theorem. *Every process X with rational spectrum is regular. For a white noise W to be the innovation of X it is necessary and sufficient that X and W be linked by the canonical* ARMA *relation* $A_P X = A_Q W$.

Proof. Let X have rational spectrum. There then exists a white noise W linked to X by the canonical ARMA relation. Since the roots of P have modulus > 1, we have seen in 3.2 that

$$(17) \qquad X_n = \sum_{k \geqslant 0} c_k W_{n-k}$$

This implies $X_m \in H_n^W$ for $m \leqslant n$ and hence $H_n^X \subset H_n^W$.

The converse inclusion is more involved since Q may have roots of modulus 1. It is a consequence of Lemma 5.2 below, after noticing that $U = A_P X$ obviously verifies $H_n^U \subset H_n^X$. Thanks to 5.2, we may then conclude that $H_n^X = H_n^W$. The process X is then regular, and W is proportional to the innovation of X. Since

$$c_0 = \frac{Q}{P}(0) = 1,$$

relation (17) shows that $W_n = X_n - p_{n-1}(X_n)$ where p_{n-1} is the orthogonal projection onto $H_{n-1}^X = H_{n-1}^W$. Thus W is exactly the innovation of X.

5.2. Lemma. *Let Y be a regular stationary process. Let Q be a polynomial having all its roots of modulus* $\geqslant 1$. *Then the relation* $U = A_Q Y$ *implies* $H_n^U = H_n^Y$ *for all n.*

Proof. Since U is a finite causal moving average we obviously have $H_n^U \subset H_n^Y$. Similarly, for any polynomial R, the process $V = A_R Y$ verifies $H_n^V \subset H_n^Y$ and letting $n \to -\infty$, we see that V must be regular.

Write $Q(z) = (1 - uz)R(z)$ with $|u| \leqslant 1$. Then the regular process $V = A_R Y$ verifies

$$V_n - uV_{n-1} = U_n$$

which by an elementary computation implies

$$V_n - u^{k+1}V_{n-k-1} = \sum_{j=0}^{k} u^j U_{n-j}, \qquad k \geq 1,$$

and in particular $(V_n - u^{k+1}V_{n-k-1})$ belongs to H_n^U for all $k \geq 1$. Let $\Phi \in L^2(\Omega,P)$ be an arbitrary r.v. Call p_j^V the orthogonal projection on H_j^V. We have

$$<\Phi, u^{k+1}V_{n-k-1}> = \bar{u}^{k+1} <p_{n-k-1}^V(\Phi), V_{n-k-1}>$$

and the modulus of this scalar product is hence bounded by $\alpha \| p_{n-k-1}^V(\Phi) \|$ where α^2 is the variance of V. Since V is regular, we have (cf. Chapter 4)

$$\lim_{j \to -\infty} p_j^V(\Phi) = 0$$

and hence the sequence $u^{k+1}V_{n-k-1}$ converges **weakly** (cf. Appendix) to 0 as $k \to +\infty$. Consequently, V_n belongs to the **weak** closure of H_n^U. But H_n^U, being a strongly closed **subspace** of $L^2(\Omega,P)$, must also be weakly closed (cf. Appendix), whence $V_n \in H_n^U$. For $m \leq n$ we then have $V_m \in H_m^U \subset H_n^U$ and hence $H_n^V = H_n^U$ since the inclusion $H_n^U \subset H_n^V$ is trivial.

To conclude the proof, one proceeds by induction on degree (Q), starting anew with the equation $V = A_R Y$ and factorizing R, etc. .

Bibliographical Hints

A widely known text on ARMA processes is *Box* and *Jenkins*' book, in which many problems dealing with the various ARMA representations of the same process, or regularity questions are only hinted at. Processes with rational spectrum are studied in detail by *Rozanov*, who does not limit himself to the ARMA point of view. Most other classical texts on time series, for instance, *Hannan*, *Priestley*, and *T. W. Anderson*, include a few chapters on processes with rational spectrum.

Chapter IX
NONSTATIONARY ARMA PROCESSES AND FORECASTING

1. Nonstationary ARMA Models

To modelize a nonstationary random phenomenon, it is tempting to write $Y_n = f(n) + X_n$ where X is stationary and the trend $f(n)$ is a deterministic function to be adequately selected; $f(n)$ is often assumed to be the sum of a polynomial in n and of linear combinations of $\cos n\lambda_j$ and $\sin n\lambda_j$, $j = 1,2, \ldots, K$. A first approach is to estimate separately f (for instance by least squares methods (cf. [15]) and the covariance structure of X.

Another method avoids the direct estimation of f by applying simple filters to Y in order to eliminate f. As seen in Chapter 7, Section 1, if P is an adequate polynomial, the image $A_P Y$ of Y by the filter A_P coincides with $A_P X$ and hence is stationary.

The point of view adopted by Box-Jenkins (cf. [10]) and others is to ignore the very existence of a trend and to consider processes Y, not specified a priori, such that $A_P Y$ is stationary and of type ARMA. These models are directly used for forecasting, without identifying the structure of Y. We are going to present the mathematical basis of this approach.

1.1. The General Nonstationary ARMA Equation

Let $Y = (Y_n)_{n \in \mathbb{Z}}$ be a discrete time process *which is not assumed to be stationary.* Let

$$P(z) = \sum_{k=0}^{p} a_k z^k$$

be an arbitrary polynomial. We *define* the process $U = A_P Y$ by

$$U_n = \sum_{k=0}^{p} a_k Y_{n-k}.$$

It is obvious that $A_P A_Q Y = A_{PQ} Y$ for all P, Q, Y.

Let us recall a few classical results for the recursion equation $A_P f \equiv 0$ where f is **deterministic** and $a_0 \neq 0$, which may be written

(1) $$\sum_{k=0}^{p} a_k f(n-k) = 0 \quad \text{for} \quad n \in \mathbb{Z}.$$

To every root τ, with multiplicity r, of the equation $P(1/\tau) = 0$ are associated r **deterministic** solutions $\varphi_1 \ldots \varphi_r$ of (1) defined by

$$\varphi_1(n) = \tau^n, \quad \varphi_2(n) = n\tau^n, \ldots, \varphi_r(n) = n^{r-1}\tau^n.$$

When τ runs through the set of inverses of all the roots of P, this supplies the **standard solutions** $f_1 \ldots f_p$ of (1), which form a *basis* for the vector space of all solutions of (1).

Given arbitrary $u_{j+1} \ldots u_{j+p}$ in $\mathbb{C}$, with j fixed in $\mathbb{Z}$, there exists a unique solution f of (1) such that

$$f(j+1) = u_{j+1}, \ldots, f(j+p) = u_{j+p}.$$

Now consider an arbitrary given function $g: \mathbb{Z} \to \mathbb{C}$, and the equation in f

(2) $$\sum_{k=0}^{p} a_k f(n-k) = g(n) \quad \text{for} \quad n \in \mathbb{Z}.$$

We may choose arbitrarily $f(0) \ldots f(p)$ and then compute uniquely all the other values of f by an obvious recursive use of (2). The space of solutions of (2) for given g is affine with dimension p and the general solution of (2) is given by $f = \tilde{g} + c_1 f_1 + \ldots + c_p f_p$, where $\tilde{g}$ is a particular solution which we are going to compute explicitly and $c_1 \ldots c_p$ are arbitrary constants.

Call δ the Dirac function defined here by $\delta(0) = 1$ and $\delta(n) = 0$ for $n \neq 0$. The convolution $\varphi * \psi$ being defined by

$$\varphi * \psi(n) = \sum_{j \in \mathbf{Z}} \varphi(j)\psi(n-j),$$

we have $A_p(\varphi * \psi) = (A_p\varphi) * \psi$ *whenever the series converge* and $\delta * \psi \equiv \psi$, so that if φ is a solution of $A_p\varphi = \delta$, then $\varphi * g = f$ will be a solution of $A_p f = g$.

Call then ψ^+ and ψ^- the unique solutions of

$$(3) \qquad A_p\psi^+ = \delta, \text{ and } \psi^+(n) = 0 \quad \text{for } n \leqslant -1$$

$$(3) \qquad A_p\psi^- = \delta, \text{ and } \psi^-(n) = 0 \quad \text{for } n \geqslant 1.$$

Let us point out that ψ^+, ψ^- can be computed recursively, and that one has $\psi^+ \equiv \chi_1 f_1 + \ldots + \chi_p f_p$ for $n \geqslant -(p-1)$, where the numbers $\chi_1 \ldots \chi_p$ are determined by the linear system $\psi^+(0) = 1/a_0$ and $\psi^+(j) = 0$ for $-1 \geqslant j \geqslant -(p-1)$. Analogous formula for ψ^-. **We shall call ψ^+ and ψ^- the two elementary inverses of A_p.**

Let $g^+(n) = g(n)1_{\{n \geqslant 0\}}$ and $g^-(n) = g(n)1_{\{n \leqslant 0\}}$. **Then $\psi^+ * g^+$ and $\psi^- * g^-$ are always defined, and coincide with the truncated convolution**

$$\psi^+ * g^+(n) = \sum_{j=0}^{n} \psi^+(j)g(n-j) \qquad n \geqslant 0$$

$$\psi^- * g^-(n) = \sum_{j=n+1}^{0} \psi^-(j)g(n-j) \qquad n < 0$$

and the preceding analysis shows that the function $\tilde{g}$ defined by relation (4) below, is a particular solution of (2)

$$(4) \qquad \begin{cases} \tilde{g}(n) = \sum_{j=0}^{n} \psi^+(j)g(n-j) & n \geqslant 0 \\ \tilde{g}(n) = \sum_{j=n+1}^{0} \psi^-(j)g(n-j) & n < 0. \end{cases}$$

Applying these elementary results to every trajectory $Y(\omega)$ of a process Y such that $A_pY = V$, we obtain the following lemma.

1.2. Lemma. *Let V be an arbitrary second-order process, which is not assumed to be stationary. Let P be an arbitrary polynomial of degree p, with complex coefficients a_k, $0 \leqslant k \leqslant p$. Assume $a_0 \neq 0$. Call $f_1 ... f_p$ the standard solutions of $A_P f = 0$ and ψ^+, ψ^- the two elementary inverses of A_P.*

Then a second-order process Y (with no stationarity hypothesis) is a solution of $A_P Y = V$ if and only if

$$Y_n = \widehat{V}_n + C_1 f_1(n) + ... + C_p f_p(n), \quad n \in \mathbb{Z} \tag{5}$$

where $C_1 ... C_p$ are arbitrary random variables in $L^2(\Omega, P)$ and $\widehat{V}$ is given by

$$\widehat{V}_n = \sum_{j=0}^{n} \psi^+(j) V_{n-j} \quad \text{for} \quad n \geqslant 0$$

(6)

$$\widehat{V}_n = \sum_{j=n+1}^{0} \psi^-(j) V_{n-j} \quad \text{for} \quad n < 0$$

which may be written $\widehat{V}^+ = \psi^+ * V^+$ and $\widehat{V}^- = \psi^- * V^-$ with obvious conventions.

In particular, if we take $V = A_Q W$ where Q is any polynomial and W any white noise, so that V is stationary, we get the following corollary: *The general "ARMA equation" $A_P Y = A_Q W$, where P, Q and the white noise W are given and arbitrary, admits, provided $a_0 \neq 0$, an infinity of second order solutions Y, which are a priori nonstationary.* These solutions form, in the vector space of second-order *processes* an *affine subspace of dimension p.*

By Chapter 8, Theorem 2.3, *these solutions are necessarily nonstationary unles every root of modulus* 1 *and multiplicity r for P is also a root of Q with multiplicity $\geqslant r$*, that is (cf. Chapter 8, Notation 2.1) unless the singular factor P_s of P divides Q.

When the signal factor P_s of P divides Q, the stationary solutions Y of $A_P Y = A_Q W$ form an affine space of dimension u where u is the number of distinct roots of P_s, by Theorem 2.3, Chapter 8. In particular if P has no root of modulus 1, then $u = 0$ and there is a unique stationary solution. *But there are always an infinity of nonstationary solutions, except in the trivial case where $P_s \equiv P$ and P divides Q.*

1.3. The ARIMA Models

Write $\Delta Y_n = Y_n - Y_{n-1}$. In many concrete cases the passage from Y to ΔY or to $\Delta^d Y$, $d \geq 1$, tends to decrease the estimated correlations, which is a (vague) clue of "increasing" stationarity (cf. Chapter 11). Let us hence introduce (following Box-Jenkins [10]) *the class of all processes Y such that $\Delta^d Y$ is a centered stationary process X with rational spectrum.* Such a process Y will be said to be of the **type** ARIMA(p,d,q) if X is of minimal type (p,q).

Let W be the innovation of X and $A_P X = A_Q W$ the canonical ARMA relation linking X and W. Let $\tilde{P}(z) = (1 - z)^d P(z)$. Then Y is a solution of $A_{\tilde{P}} Y = A_Q W$. In particular *if Q is not divisible by $(1 - z)^d$, Y is certainly not stationary.* By 1.1 the general solution of $\Delta^d Y = X$ is given by, for $n \geq 0$,

$$Y_n = C_0 + C_1 n + \dots + C_{d-1} n^{d-1} + \sum_{j=0}^{n} \psi_d^+(j) X_{n-j} \tag{7}$$

$$\psi_d^+(n) = \frac{(n+1)(n+2)\dots(n+d-1)}{(d-1)!} \qquad \text{for } d \geq 1,\ n \geq 0$$

$$\psi_1^+(n) \equiv 1 \qquad \text{for } d = 1,\ n \geq 0$$

(analogous formulas for $n \leq 0$), where $C_0 \dots C_{d-1}$ are arbitrary r.v. in $L^2(\Omega,P)$.

Assume that d is known and that X is ergodic. Then, given *one* infinite trajectory $\{Y_n(\omega)\}_{n \in \mathbf{Z}}$ *with ω fixed in Ω*, we may obviously determine the $X_n(\omega) = \Delta^d Y_n(\omega)$ and hence by (7) the polynomial

$$J_\omega(n) = \sum_{j=0}^{d-1} C_j(\omega) n^j.$$

This in turn determines the coefficients $C_0(\omega) \dots C_{d-1}(\omega)$ of J_ω. Since X is ergodic the infinite trajectory $X(\omega)$ completely determines the distribution of X (cf. Chapter 3, Section 4.2). But the knowledge of $C_0(\omega) \dots C_{d-1}(\omega)$ with *fixed ω supplies no* information on the $C_j(\omega')$, $\omega' \neq \omega$ and hence *no information* on the joint distribution of $(C_0, \dots, C_{d-1})$ or its moments.

Hence a model of type (7) cannot generally be identified starting from a single infinite trajectory of Y. Of course it becomes identifiable if the experiment (the observation of a long segment of the trajectory of Y) can be repeated in identical conditions, which often is not the case, e.g. in econometrics.

The situation changes if one constrains the r.v. $C_0 \dots C_{d-1}$ to be almost surely constant, which is equivalent to the statement $Y_n = J(n) + \tilde{X}_n$ where $\tilde{X} = \psi^+ * X$ for $n \geq 0$, and $J(n)$ is a deterministic polynomial. *Then Y is obviously identifiable in distribution starting from the observation of a single infinite trajectory of Y*, assuming as usual X to be ergodic.

If Y verifies (7) and if all the data available amount to the trajectory $Y(\omega)$, with *fixed* ω, we can always find Y' verifying (7) with constant $C'_0 \dots C'_{d-1}$, such that $Y'(\omega) \equiv Y(\omega)$, and hence practically indistinguishable from Y on the basis of the available observations. Since Y' has the merit of being identifiable, *it seems sensible to limit oneself to modelization by identifiable* ARIMA *processes, that is by solution of* (7) *with constant* $C_0 \dots C_{d-1}$.

1.4. The Seasonal ARMA Models

When the physical nature or the general rough pattern of a random phenomenon Y suggests the presence of intrinsic periodicities (seasonal effects in economics or meteorology, periodic signals perturbed by noise), it is natural to consider the differences $(Y_n - Y_{n-s})$ if the segments of trajectories obtained by the translation $Y_n \to Y_{n+s}$ have strong similarities.

Such a filter A_P, associated to $P(z) = (1 - z^s)$, tends to decorrelate observations presenting seasonal effects of period s. In the presence of several seasonal effects, of periods $s_1 \dots s_k$, one may even try to "**deseasonalize**" or "stationarize" Y by the filter A_R with

$$R(z) = (1 - z^{s_1})^{d_1} \dots (1 - z^{s_k})^{d_k}$$

without excluding $s_1 = 1$, where $d_1 \dots d_k$ are small integers.

To formalize an approach introduced by Box-Jenkins, we consider *the class of all processes Y (with no stationarity hypothesis) such that there exists*

$$R(z) = (1 - z^{s_1})^{d_1} \dots (1 - z^{s_k})^{d_k}$$

and a centered stationary process X with rational spectrum verifying $A_R Y = X$. We shall then say that *Y is of type* ARMA *with seasonal effects.* Clearly the ARIMA processes are a particular case of this type of model.

Let $A_P X = A_Q W$ be the canonical ARMA relation satisfied by X and its innovation W. Then Y satisfies $A_{RP} Y = A_Q W$ and hence Y *is certainly not stationary if Q has no root of modulus 1.*

Let $D = s_1 d_1 + ... + s_k d_k$. The standard solutions $f_1 ... f_D$ of $A_R f \equiv 0$ are of the type (cf. 1.1)

(8) $f(x) = x^\alpha \cos \beta x$ and $f(x) = x^\alpha \sin \beta x$

where $\beta = 2\pi j/s$, $0 \leqslant j \leqslant s - 1$, $0 \leqslant \alpha \leqslant d - 1$, (s,d) being any one of the $(s_1, d_1) ... (s_k, d_k)$.

The elementary inverses ψ^+ and ψ^- of A_R (cf. 1.1) are (precise) linear combinations of functions of type (8). By 1.1 the general solution Y of $A_{RP} Y = A_Q W$ is

$$Y_n = C_1 f_1(n) + ... + C_D f_D(n) + \widetilde{X}_n$$

where $\widetilde{X}$ is given by (6), that is, $\widetilde{X}^+ = \psi^+ * X^+$ and $\widetilde{X}^- = \psi^- * X^-$, and the C_j are *arbitrary* r.v. in $L^2(\Omega, P)$.

As in 1.3, Y *is generally not identifiable in distribution starting from an infinite fixed trajectory, even if X is ergodic.* On the other hand, for X ergodic, and *constant* $C_1 ... C_D$, Y is identifiable in distribution.

As in 1.3, we conclude that to modelize phenomenons which cannot be reproduced with "identical environments," there is no practical or theoretical advantage in the use of nonidentifiable seasonal ARMA processes Y, since for any given $\omega \in \Omega$, there exists an identifiable seasonal ARMA process Y' such that $Y'(\omega)$ is indistinguishable of $Y(\omega)$.

1.5. Multiplicative Seasonal Models

Let Y be a random phenomenon exhibiting seasonal effects of period M. For $0 \leqslant s \leqslant M - 1$ the processes Y^s defined by $Y_n^s = Y_{s+Mn}$ have, in numerous concrete cases, a probabilistic structure *approximately independent of s.* We may then attempt to modelize Y^s by an ARIMA model whose parameters are independent of s, of the type

$$A_{RP} Y^s = A_Q W^s \qquad s = 0, 1, ..., M - 1$$

where the white noise W^s may depend on s, but P, Q, R are independent of s, $R(z) = (1 - z)^D$, and P, Q are coprime, with all their roots of modulus > 1.

Let $\breve{R}(z) = R(z^M)$, $\breve{P}(z) = P(z^M)$, $\breve{Q}(z) = Q(z^M)$ and *define* a process U by $U_{s+Mn} = W^s_n$. Note that generally U is not a white noise, not even a stationary process. These definitions trivially imply

$$[A_{\breve{R}\breve{P}}Y]_{s+Mn} = [A_{RP}Y^s]_n \quad \text{for } n \in \mathbb{Z}\,,\ 0 \leqslant s \leqslant M-1$$

$$[A_{\breve{Q}}U]_{s+Mn} = [A_Q W^s]_n \quad \text{for } n \in \mathbb{Z}\,,\ 0 \leqslant s \leqslant M-1$$

whence finally

$$(9) \qquad A_{\breve{R}\breve{P}}Y = A_{\breve{Q}}U.$$

Let us attempt to modelize U by an ARIMA process. Assume then the existence of polynomials π, χ, ρ and of a white noise W such that

$$(10) \qquad A_{\rho\pi}U = A_{\chi}W$$

with $\rho(z) = (1 - z)^d$, and where π, χ are coprime and have all their roots of modulus > 1.

By (9) and (10) we get

$$(11) \qquad A_{\rho\breve{R}\pi\breve{P}}Y = A_{\chi\breve{Q}}W.$$

Contrary to what one could believe at first glance, this is not enough to guarantee the stationarity of $A_{\rho\breve{R}}Y$. But *we may certainly impose*, to the model which we are building, *the extra hypothesis of stationarity for* $A_{\rho\breve{R}}Y$.

Since $\pi\breve{P}$ has no root of modulus 1, there exists a unique stationary process X verifying

$$(12) \qquad A_{\pi\breve{P}}X = A_{\chi\breve{Q}}W$$

and X must then have rational spectrum (cf. Chapter 8, Theorem 2.3). We have then, by (11), (12), and the uniqueness of X, $A_{\rho\breve{R}}Y = X$. Since

$$\rho(z)\breve{R}(z) = (1 - z)^d(1 - z^M)^D,$$

we see that *under these hypotheses, Y is an* ARMA *process with seasonal effects.*

We shall say *that Y is of* ARMA *type with multiplicative seasonal effects*, if Y verifies (11), with $A_{\rho R}Y$ assumed to be stationary. This type of decomposition (introduced by Box-Jenkins) permits, when the model fits the data, *the study of periodic effects with long memory, using a small number of parameters.* For we may have P, Q, π, χ of *small degrees*, while the canonical ARMA relation (12) of X has *high order*. For instance, in the frequent case where $M = 12$ (monthly observations), if we take P, Q, π, ψ of degree 1, then $P\pi$ and $Q\chi$ are of degree 13. The a priori modelization of X by an ARMA(13,13) would involve the computation of 27 parmeters, while the multiplicative model, if adequate, depends only on 5 parameters.

2. Linear Forecasting and Processes with Rational Spectrum

For a stationary process X with rational spectrum, it is easy to explicitly compute the optimal (linear) predictor of X_{n+j} given the X_k, $k \leq n$. As in Chapter 4 we shall denote it $\hat{X}_{n+j} = p_n(X_{n+j})$ where p_n is the projection on H_n^X.

2.1. Theorem. *Let X be a centered stationary process with rational spectrum. Let*

$$f(\lambda) = \frac{\sigma^2}{2\pi}\left|\frac{Q}{P}(e^{-i\lambda})\right|^2$$

be the canonical Fejer-Riesz representation of its spectral density, with

$$\frac{Q}{P}(0) = 1.$$

Let

$$\frac{Q}{P}(z) = \sum_{k\geq 0} c_k z^k$$

be the Laurent expansion of Q/P for $|z| \leq 1$. Fix $j \geq 1$. Define g_j by

$$g_j(\lambda) = e^{ij\lambda}\left[1 - \left(\sum_{k=0}^{j-1} c_k e^{-ik\lambda}\right)\frac{P}{Q}(e^{-i\lambda})\right]. \tag{13}$$

The optimal linear predictors $\hat{X}_{n+j} = p_n(X_{n+j})$ *define, when* n *runs through* $\mathbb{Z}$, *a stationary process* $\hat{X}^j$, *and* $\hat{X}^j = F_{g_j}X$ *is the image of* X *by the filter with response function* g_j. *Moreover the* j *steps-prediction error is given by*

$$(14) \qquad E|\hat{X}_{n+j} - X_{n+j}|^2 = \sigma^2 \sum_{k=0}^{j-1} |c_k|^2 .$$

Proof. By Chapter 8, Theorem 5.1 we have

$$X_n = \sum_{k \geq 0} c_k W_{n-k},$$

where W is the innovation of X and has variance σ^2, and where the c_k are as above. In particular

$$(15) \qquad X_{n+j} = \sum_{k \geq j} c_k W_{n+j-k} + \sum_{0 \leq k \leq j-1} c_k W_{n+j-k} \, .$$

The first sum S is in H_n^W and the second one is orthogonal to H_n^W. Since $H_n^X = H_n^W$ we must then have

$$(16) \qquad \hat{X}_{n+j} = p_n(X_{n+j}) = S = \sum_{k \geq 0} c_{k+j} W_{n-k}$$

which may be written $\hat{X}^j = F_h W$, where F_h is the filter with response function

$$h(\lambda) = \sum_{k \geq 0} c_{k+j} e^{-ik\lambda} = e^{ij\lambda}\left[\frac{Q}{P}(e^{-i\lambda}) - \sum_{k=0}^{j-1} c_k e^{-ik\lambda}\right].$$

Since $W = F_\varphi A_P X$ with $\varphi = [1/h_Q]$ (notations from Chapter 7, Section 3.6), we obtain $\hat{X}^j = F_h F_\varphi A_P X = F_g X$ with $g = h[1/h_Q]h_P$, whence

$$g(\lambda) = h(\lambda)\,\frac{P}{Q}(e^{-i\lambda})$$

which proves (13).

By (15) and (16) we have

$$X_{n+j} - \hat{X}_{n+j} = \sum_{k=0}^{j-1} c_k W_{n+j-k}$$

which proves (14).

Example. Let us compute $\hat{X}_{n+1}$ explicitly *when* Q *has no root of modulus* 1. Then the roots of Q have moduli > 1, and

$$\frac{P}{Q}(z) = \sum_{k\geq 0} d_k z^k \quad \text{for } |z| \leq 1.$$

Moreover

$$c_0 = \frac{Q}{P}(0) = 1 \quad \text{and} \quad d_0 = \frac{P}{Q}(0) = 1.$$

Formula (13) yields

$$g_1(\lambda) = -\sum_{k\geq 0} d_{k+1} e^{-ik\lambda} \quad \text{with} \quad \sum_{k\geq 0} |d_k| < \infty$$

whence

(17) $$\hat{X}_{n+1} = -\sum_{k\geq 0} d_{k+1} X_{n-k}\,.$$

In particular, *if X is autoregressive*, $Q(z)$ is identically 1 and (17) implies, with the notations $P(z) = \Sigma_{k=0}^{p} a_k z^k$ and $a_0 = P(0) = 1$,

$$a_0 \hat{X}_{n+1} + a_1 X_n + a_2 X_{n-1} + \dots + a_p X_{n-p+1} = 0$$

which justifies the term "autoregressive."

In the general case, the practical computation of predictors is simplified by the use of recursive relations which are easy consequences of the canonical ARMA relation.

2.2. Recursion Relations Between Predictors

Let

$$P(z) = \sum_{k=0}^{p} a_k z^k \quad \text{and} \quad Q(z) = \sum_{\ell=0}^{q} b_\ell z^\ell.$$

Let X be as above, and call W the innovation of X. Write the canonical ARMA relation

$$\sum_{k=0}^{p} a_k X_{n-k} = \sum_{\ell=0}^{q} b_\ell W_{n-\ell} \qquad n \in \mathbb{Z}.$$

Write this relation at time $(n + m)$ and project it orthogonally on $H_n^X = H_n^W$. The projection of X_k on H_n^X is X_k if $k \leq n$, and coincides when $k \geq n + 1$, with the optimal predictor of X_k given H_n^X, which we denote $p_n(X_k)$. The projection of W_k on H_n^X is W_k if $k \leq n$, and zero if $k \geq n + 1$. Whence the following formulas, true for all $n \in \mathbb{Z}$,

(18) $$\sum_{k=0}^{p} a_k p_n(X_{n+m-k}) = 0, \qquad m \geq q + 1$$

(19) $$\sum_{k=0}^{p} a_k p_n(X_{n+m-k}) = \sum_{\ell=m}^{q} b_\ell W_{n+m-\ell}, \quad 0 \leq m \leq q$$

with $p_n(X_{n+m-k}) = X_{n+m-k}$ if $m - k \leq 0$.

2.3. Practical Use of Recursions (18) and (19)

Assume that $\hat{X}_{n+1} = p_n(X_{n+1})$ has already been computed, for instance by (17). By definition (cf. Chapter 4, Section 3) the innovation W verifies

$$W_k = X_k - p_{k-1}(X_k) \qquad \text{for all } k$$

which immediately yields the W_k for $k \leq n$. This provides the values of all right-hand sides in (19), since they only involve the W_k, $n - q \leq k \leq n$. We may then use (19) with m increasing from 2 to q, to successively compute $p_n(X_{n+2}) \dots p_n(X_{n+q})$. Then thanks to (18) with m increasing from $q + 1$ to infinity, we successively compute $p_n(X_{n+q+1}) \dots p_n(X_{n+m})$ and so on.

2.4. Behaviour of Long-Term Forecasts

Assume the $X_k(\omega)$, $k \leq n$ to be already observed, and fixed. If at time n we forecast the whole future X_{n+m}, $m \geq 1$, we obtain, for fixed n and ω, the function $\varphi(m) = p_n(X_{n+m})$ which by (18) verifies

(20) $$\sum_{k=0}^{p} a_k \varphi(m - k) = 0 \qquad \text{for } m \geq q + 1.$$

Hence (cf. 1.1) φ is a linear combination of the standard solutions $f_1 \dots f_p$ of (20), which are all of type $m \to m^\alpha \tau^m$, $0 \leq \alpha \leq r - 1$, where τ is a root with multiplicity r of $P(1/\tau) = 0$. In particular, since $A_P X = A_Q W$ is the canonical ARMA relation of X, we have $|\tau| < 1$ and we see that $\varphi(m)$ tends to zero at exponential speed when $m \to +\infty$. Thus the distant X_{n+m} are estimated by a value $\varphi(m)$ close to $0 = E(X_{n+m})$. The information supplies by H_n^X has little or no long term impact.

2.5. Updating of Forecasts

Assume that, at time n, one has computed a large number of predictors $\varphi(m) = p_n(X_{n+m})$. As soon as the observation X_{n+1} becomes available it is natural to readjust the forecasts of $X_{n+2} \ldots X_{n+m}$. The problem here is to compute $\psi(m) = p_{n+1}(X_{n+m})$ starting with φ and X_{n+1}. Formula (16) immediately implies

$$p_n(X_{n+m}) = c_m W_n + c_{m+1} W_{n-1} + \cdots$$

whence $\psi(m) = \varphi(m) + c_{m-1} W_{n+1}$ and the formula

$$\psi(m) = \varphi(m) + c_{m-1}[X_{n+1} - p_n(X_{n+1})]$$

valid for $m \geqslant 2$ which links the new forecasts ψ to the former ones φ.

3. Time Inversion and Estimation of Past Observations

3.1. Time Inversion

Let X be a real valued stationary process with rational spectrum. Let

$$f(\lambda) = \frac{\sigma^2}{2\pi} \left| \frac{Q}{P}(e^{-i\lambda}) \right|^2$$

be the canonical form of its spectral density, and call W the innovation of X.

The process Y defined by $Y_n = X_{-n}$ is still stationary and its spectral measure μ^Y is symmetric of the spectral measure μ^X (cf. Chapter 6, Section 3.4). Since X is real valued, μ^X is invariant by the natural symmetry of the torus, and hence $\mu^X = \mu^Y$. In particular Y has the same spectral density f as X, and thus has rational spectrum. Since f is in canonical form, the canonical ARMA relation verified by Y may be written $A_P Y = A_Q \tilde{W}$ where $\tilde{W}$ is the innovation of Y and must have *the same variance* σ^2 *as* W.

Write $U_n = \tilde{W}_{-n}$. Clearly *U is a white noise with variance* σ^2. For a process X, called (linear) **future** at time n, the (closed) linear envelope F_n^X of the X_k, $k \geqslant n$. Since Y and $\tilde{W}$ have the same part at time n, *X and U clearly have the same future*

$F_n^X = F_n^U$ *at time n*, for all $n \in \mathbb{Z}$.

If

$$P(z) = \sum_{k=0}^{p} a_k z^k, \quad Q(z) = \sum_{\ell=0}^{q} b_\ell z^\ell,$$

the relation $A_P Y = A_Q \widetilde{W}$ gives

$$\sum_{k=0}^{p} a_k Y_{n-k} = \sum_{\ell=0}^{q} b_\ell \widetilde{W}_{n-\ell}\,.$$

Replacing n by $-n$, and then Y_{-n-k}, $\widetilde{W}_{-n-\ell}$ by X_{n+k}, $U_{n+\ell}$, we obtain

(21) $$\sum_{k=0}^{p} a_k X_{n+k} = \sum_{\ell=0}^{q} b_\ell U_{n+\ell} \qquad n \in \mathbb{Z}.$$

3.2. Estimation of Past Values

Assume that the X_k, $k \geq n$ are known. We want to estimate "missing" observations X_{n-1}, X_{n-2}, ... Let Y be the process deduced from X by time inversion, as in 3.1. Call p_m^Y the projection on H_m^Y. By Theorem 2.1, the process $\hat{Y}^j$ is defined for *fixed* j by $\hat{Y}_m^j = p_m^Y(Y_{m+j})$ may be written $\hat{Y}^j = F_{g_j} Y$, with g_j given by (13), since the canonical ARMA relation of Y has the same coefficients as the canonical relation of X.

This implies, with the notation $f_m(\lambda) = e^{im\lambda}$,

$$\hat{Y}_m^j = \int f_m g_j \, dZ^Y, \qquad m \in \mathbb{Z}.$$

But (cf. Chapter 6, Section 3.4), if s: $\mathbb{T} \to \mathbb{T}$ is the (mod 2π) symmetry, we have seen that

$$\int \varphi \, dZ^Y = \int \varphi \circ s \, dZ^X,$$

whence

$$\hat{Y}_m^j = \int f_{-m}(\lambda) g_j(-\lambda) dZ^X(\lambda), \quad m \in \mathbb{Z}.$$

Setting $m = -n$ and $h_j(\lambda) = g_j(-\lambda)$ we see that

$$\hat{Y}_{-n}^j = \int f_n h_j \, dZ^X = S_n, \quad n \in \mathbb{Z}$$

where $S = F_{h_j} X$ is the image of X by the filter with response function h_j. But by definition of Y, $\hat{Y}^j_{-n}$ is precisely the projection of X_{n-j} onto the future F^X_n.

Finally, *the optimal (linear) estimate of* X_{n-j} *given the* X_k, $k \geq n$, *is the image (at time n) of X by the filter with response function* $h_j(\lambda) = g_j(-\lambda)$ *where* g_j *is given by* (13).

4. Forecasting and Nonstationary ARMA Processes

4.1. Computation of Predictors

Let Y be an ARMA type process with seasonal effects (cf. 1.4). Thus there exists

$$R(z) = (1 - z^{s_1})^{d_1} ... (1 - z^{s_k})^{d_k}$$

such that $X = A_R Y$ is centered stationary with rational spectrum. Let $D = d_1 s_1 + .. + d_k s_k$. We use the explicit representation (cf. 1.1, 1.4) of Y

$$(22) \qquad Y = C_1 f_1 + ... + C_D f_D + \tilde{X}$$

where $f_1 ... f_D$ are the standard solutions of $A_R f = 0$ and

$$\tilde{X}_n = \sum_{j=0}^{n} \psi^+(j) X_{n-j} \quad \text{for } n \geq 0$$

(with an analogous formula for $n \leq 0$), and where ψ^+, ψ^- are the elementary inverses of A_R. The terms $C_1 ... C_D$ are arbitrary r.v. *which we shall assume to be uncorrelated with the* X_m, $m \in \mathbb{Z}$, or, in other words, to be orthogonal to H^X. This hypothesis is of course satisfied when $C_1 ... C_D$ are constants, i.e. in the identifiable case.

It is then easily seen that for $n \geq 0$ we have $H^Y_n = H^X_n \oplus F$ where F is the vector space generated by $C_1 ... C_D$. For $n < 0$ the result is more involved, with $H^Y_n = H^X_n \oplus F_n$ where F_n varies with n but remains of finite dimension.

We shall solve the simple case where $n \geq 0$. Since F and H^X are assumed to be orthogonal, we have $p^Y_n(X_m) = p^X_n(X_m)$ for all $m,n \geq 0$, where p^Y_n, p^X_n are the projections on H^Y_n, H^X_n.

Write

$$R(z) = \sum_{k=0}^{D} u_k z^k$$

and

$$\hat{Y}_m = p_m^Y(Y_m), \quad \hat{X}_m = p_n^Y(X_m) = p_n^X(X_m).$$

By Section 2, the $\hat{X}_m$ can be computed explicitly. By projection the equality $A_R Y = X$ yields

$$\sum_{k=0}^{D} u_k \hat{Y}_{m-k} = \hat{X}_m \qquad m \geqslant 0 \tag{23}$$

and since $\hat{Y}_m = Y_m$ for $m \leqslant n$, this supplies by recursion all the predictors $\hat{Y}_m$, $m \geqslant 0$.

For $n < 0$ the recursion still holds with $\hat{X}_m = p_n^Y(X_m)$, but we do not have $p_n^Y(X_m) = p_n^X(X_m)$ any more. The computations are of course feasible but the results are inelegant.

4.2. Error of Prediction

By projection, formula (22) implies, for $m \geqslant 0$,

$$\hat{Y}_m = \sum_{j=1}^{D} C_j f_j(m) + \sum_{j=0}^{m} \psi^+(j)\hat{X}_{m-j}$$

whence for $k,n \geqslant 0$

$$Y_{n+k} - \hat{Y}_{n+k} = \sum_{\ell=1}^{k} \psi^+(k-\ell)(X_{n+\ell} - \hat{X}_{n+\ell}).$$

But we have seen (cf. Section 2) that, with

$$\frac{Q}{P}(z) = \sum_{k \geqslant 0} c_k z^k,$$

$$X_{n+\ell} - \hat{X}_{n+\ell} = c_0 W_{n+\ell} + \dots + c_{\ell-1} W_{n+1}.$$

With the notation

$$\eta_r = \sum_{j=0}^{r} \psi^+(r-j)c_j$$

this yields

$$Y_{n+k} - \hat{Y}_{n+k} = \sum_{j=0}^{k-1} \eta_j W_{n+k-j}$$

and the k-steps error of prediction, for $n \geqslant 0$,

$$E|Y_{n+k} - \hat{Y}_{n+k}|^2 = \sigma^2 \sum_{j=0}^{k-1} \eta_j^2 .$$

Let us point out that for $n \geqslant 0$, the error of prediction does not depend on n for $n \geqslant 0$, although Y is nonstationary. The result is different for $n < 0$.

4.3. Long-Term Behaviour of Predictors

Since $H^X = H^W$, the hypothesis 4.2 implies $p_n^Y(W_m) = p_n^W(W_m)$ for $m,n \geqslant 0$. The canonical ARMA relation $A_P X = A_Q W$ and relation (23), which may be written $A_R \hat{Y} = \hat{W}$ yield, with the notation $\hat{W}_m = p_n^W(W_m) = p_n^Y(W_m)$, the following recursion

$$A_{RP}\hat{Y} = A_Q\hat{W} \quad \text{for} \quad m \geqslant q = \text{degree}(Q)$$

whence $A_{RP}\hat{Y} = 0$ for $m \geqslant n + q + 1$ since $\hat{W}_j = 0$ when $j \geqslant n+1$. This allows the recursive computation of $\hat{Y}_m$ as soon as $\hat{Y}_{m-1}, \hat{Y}_{m-2}, \ldots$ are known, for $m \geqslant n + q + 1$.

In particular for $m \geqslant n + q + 1$ the function $\varphi(m) = \hat{Y}_m$ is a linear combination (with coefficients depending on the observations Y_k, $k \leqslant n$) of the standard solutions $g_1 \ldots g_{D+p}$ of $A_{RP}g = 0$. Since the roots of P have modulus > 1, the standard solutions of $A_P g = 0$ tend exponentially fast to 0 as $m \to +\infty$, and the *long-term* behaviour of $\varphi(m)$ is hence that of a linear combination of standard solutions $f_1 \ldots f_D$ of $A_R f = 0$.

For R, P of small degrees it is a tedious, but easy task to classify the various types of graphs for φ. This botanic of prediction curves is sometimes used to select a priori a plausible seasonal model, the principle being to favor models having a typical prediction curve which "looks like" the graph of observations.

Bibliographical Hints

The presentation given here for nonstationary processes and seasonal phenomena has its origins in econometric works. Part of the chapter formalizes mathematically several points of view studied by *Box* and *Jenkins*. Their work has indeed strongly influenced the practical use of time series; it is centered on forecasting problems with a wider scope than econometrics (including, in particular, the control of industrial processes). Other quite empirical formulations of filtering and seasonality are widely used in econometrics: see, for instance, *Montgomery* and *Johnson*, the beginning of *Anderson's* book, or some chapters of *Gourieroux-Montfort*. The problem of seasonality also appears in biology for the study of rhythms.

Chapter X
EMPIRICAL ESTIMATORS AND PERIODOGRAMS

1. Empirical Estimation

1.1. Mean and Covariances

Let X be a stationary process. Given the observations $X_1 \ldots X_{N+k}$ we seek to estimate $\theta = E[f(X_1 \ldots X_k)]$ where $f\colon \mathbb{R}^k \to \mathbb{R}$ is known. *If X is strictly stationary and ergodic*, it is natural to estimate θ by

$$\hat{\theta}_N = \frac{1}{N} \sum_{j=1}^{N} f(X_{j+1} \ldots X_{j+k}),$$

which has the merit of being *a consistent estimator*, that is $\lim_{N\to\infty} \hat{\theta}_N = \theta$, P-almost surely.

This leads us to estimate *the mean* $M = EX_n$ by

$$\bar{X}_n = \frac{1}{N}(X_1 + \ldots + X_N)$$

and the covariance $r_k = E[(X_n - M)(X_{n+k} - M)]$ by

$$\bar{r}_k(N) = \frac{1}{N} \sum_{j=1}^{N} X_j X_{j+k} - \bar{X}_N^2$$

when M is unknown, and by

$$\bar{r}_k(N) = \frac{1}{N} \sum_{j=1}^{N} X_j X_{j+k} - M^2$$

when M is known.

Even when X is only second-order stationary, the estimators $\overline{X}_N$ and $\overline{r}_k(N)$ are often used in practice. Indeed one has $E[\overline{r}_k(N)] = r_k$ and $E(\overline{X}_N) = M$; moreover, the convergence in $L^2(\Omega,P)$ of $\overline{r}_k(N)$ and $\overline{X}_N$ towards r_k and M holds under "weak" assumptions. Let us indicate without proof two such general results, which by the way will not be used in the sequel.

1.1. Proposition. *Let X be second-order stationary, with mean M. For the empirical mean $\overline{X}_N$ to converge towards M in $L^2(\Omega,P)$ it is necessary and sufficient that the spectral measure μ of X verifies $\mu\{0\} = 0$. If moreover X has a spectral density f continuous at* 0, *one has*

$$\lim_{N\to\infty} NE[(\overline{X}_N - M)^2] = 2\pi f(0).$$

Proposition. *Let X be a second-order stationary process. Assume that $E[X_m^4]$ remains bounded for $m \in \mathbb{Z}$ and that*

$$\lim_{|m-n|\to\infty}[E(X_m X_{m+k} X_n X_{n+k}) - E(X_m X_{m+k}) E(X_n X_{n+k})] = 0.$$

Then

$$\lim_{N\to\infty} E\{[\overline{r}_k(N) - r_k]^2\} = 0.$$

1.2. Empirical Estimators and Asymptotic Distributions

From the point of view of linear prediction, a second-order stationary process is identified as soon as one knows its mean M and its covariances r_k. In this context, it is natural to restrict oneself to the estimation of parameters $\theta \in \mathbb{R}^k$ of the form $\theta = f(M, r_0 \ldots r_j \ldots)$ where f is a function from $\mathbb{R}^{\mathbb{N}}$ to $\mathbb{R}^k$. **The empirical estimator** of θ is then by definition

$$\overline{\theta}_n = f(\overline{X}_n, \overline{r}_0(n), \ldots \overline{r}_j(n) \ldots).$$

For instance *if f is continuous, $\overline{\theta}_n$ is a consistent estimator of θ as soon as X is ergodic.*

From a practical point of view, it is useful to obtain approximations for the distribution of $(\overline{\theta}_n - \theta)$ when n is large. When X is gausian we shall see that the distribution of $\sqrt{n}(\overline{\theta}_n - \theta)$ converges tightly toward a centered gaussian distribution.

Recall that a sequence μ_n of finite positive measures on $\mathbb{R}^k$ **converges tightly** toward a bounded measure μ if

$$\lim_{n\to\infty} \int_{\mathbb{R}^k} g\, d\mu_m = \int_{\mathbb{R}^k} g\, d\mu$$

for every bounded continuous function g: $\mathbb{R}^k \to \mathbb{R}$. By *Levy's theorem* (cf. Appendix), this is equivalent to checking that the Fourier transforms $\hat{\mu}_n(v)$ converge toward $\hat{\mu}(v)$ for all v in an open neighborhood of 0.

2. Periodograms

2.1. An "Empirical Estimator" of the Spectral Density

Let X be a centered, real-valued, stationary second-order process. Set $r_k = E(X_n X_{n+k})$ and

(1) $$\hat{r}_k(n) = \frac{1}{n} \sum_{j=1}^{n-|k|} X_j X_{j+k} \qquad \text{for } |k| \leqslant n-1.$$

Note that $E[\hat{r}_k(n)] = (1 - (|k|/n))r_k$ and that for $n \to +\infty$ the $\hat{r}_k$ are equivalent to the $\bar{r}_k$ considered above.

If X has spectral density f, the expansion

$$2\pi f(\lambda) = \sum_k r_k e^{-ik\lambda} = r_0 + \sum_{k\geqslant 1} r_k(e^{-ik\lambda} + e^{ik\lambda})$$

suggest the construction of an "estimator" $I_n(\lambda)$ for $f(\lambda)$ by

(2) $$2\pi I_n(\lambda) = \hat{r}_0(n) + \sum_{k=1}^{n-1} \hat{r}_k(n)(e^{-ik\lambda} + e^{ik\lambda})$$

where $\lambda \in \mathbb{T} = [-\pi,\pi[$. Using (1), this may be rewritten as

(3) $$2\pi I_n(\lambda) = \frac{1}{n} \left| \sum_{k=1}^{n} X_k e^{-ik\lambda} \right|^2 \geqslant 0.$$

The function I_n: $\mathbb{T} \to \mathbb{R}^+$ is called **the periodogram of order** n of X. It is a *random element* of the space $C(\mathbb{T})$ of bounded continuous functions on $\mathbb{T}$. By (2) one always has

(4) $$2\pi E[I_n(\lambda)] = r_0 + \sum_{1\leqslant|k|\leqslant n-1} \left(1 - \frac{|k|}{n}\right) r_k(e^{-ik\lambda} + e^{ik\lambda}).$$

In particular *if* $\Sigma_k |r_k|$ *is finite* the dominated convergence theorem yields

$$\lim_{n\to+\infty} E[I_n(\lambda)] = f(\lambda), \qquad \lambda \in \mathbb{T}.$$

Nevertheless, $I_n(\lambda)$ is a *very bad pointwise estimator of* $f(\lambda)$ since for fixed λ, $E([I_n(\lambda) - f(\lambda)]^2)$ does not generally converge to zero when $n \to \infty$. For instance, if ε is a stationary gaussian white noise, with periodogram I_n,

$$E\left[I_n(\lambda) - \frac{\sigma^2}{2\pi}\right]^2 = \sigma^4\left[1 - \frac{1}{n} + \frac{2}{n^2}\left|\frac{1 - e^{-i\lambda}}{1 - e^{-in\lambda}}\right|^2\right]$$

where $\sigma^2 = E\varepsilon_n^2$, and thus $E[I_n(\lambda) - (\sigma^2/2\pi)]^2$ does not converge toward zero as $n \to +\infty$, for fixed λ. On the other hand, the integrals in λ of $I_n(\lambda)$ are sensible estimates of the corresponding integrals of $f(\lambda)$, as we shall presently see.

More generally, call μ the spectral measure of X and consider the following random variables, where φ is an arbitrary bounded function on $\mathbb{T}$,

(5) $$I_n(\varphi) = \int_{\mathbb{T}} \varphi(\lambda) I_n(\lambda) d\lambda$$

and introduce the functional

(6) $$I(\varphi) = \int_{\mathbb{T}} \varphi(\lambda) d\mu(\lambda)$$

which is equal to $\int_{\mathbb{T}} \varphi(\lambda) f(\lambda) d\lambda$ when X has spectral density f.

2.2. Theorem. *If the process X, defined on (Ω,P) is centered, strictly stationary and ergodic, then P-almost surely, one has*

$$\lim_{n\to\infty} I_n(\varphi) = I(\varphi) \qquad \textit{for all } \varphi \in C(\mathbb{T}).$$

Proof. Let μ_n be the (random) positive measure defined by $d\mu_n/d\lambda = I_n$. The point is to prove that P-almost surely, μ_n converges tightly towards μ. By Levy's theorem (see Appendix) this is equivalent to checking that for every integer k, one has P-a.s.

$$\lim_{n\to\infty} \int_{\mathbb{T}} e^{ik\lambda} d\mu_n(\lambda) = \int_{\mathbb{T}} e^{ik\lambda} d\mu(\lambda).$$

But the right-hand side is equal to r_k, and by (2) the left-hand side is $\lim_{n\to\infty} \hat{r}_k(n)$. The ergodic theorem (cf. Chapter 3, Theorem 4.2) concludes the proof.

2.3. Toeplitz Matrices and Periodogram

Let φ: $\mathbb{T} \to \mathbb{R}$ be a (Lebesgue) square integrable function. Call $\hat{\varphi}_k$ the kth Fourier coefficient of φ

$$\hat{\varphi}_k = \frac{1}{2\pi} \int_{\mathbb{T}} e^{ik\lambda} \varphi(\lambda) d\lambda.$$

Define the **Toeplitz matrix of order *n* for φ** by

$$(7) \qquad T_n(\varphi) = [\hat{\varphi}_{j-k}]_{1 \leqslant j,k \leqslant n} \, .$$

For

$$x = \begin{bmatrix} x_1 \\ \vdots \\ x_n \end{bmatrix} \in \mathbb{C}^n$$

a simple computation yields

$$(8) \qquad x^* T_n(\varphi) x = \frac{1}{2\pi} \int_{\mathbb{T}} \left| \sum_{1 \leqslant k \leqslant n} x_k e^{-ik\lambda} \right|^2 \varphi(\lambda) d\lambda.$$

The matrix $T_n(\varphi)$ which is obviously *hermitian*, is hence *positive definite whenever* $\varphi \geqslant 0$, *provided* φ *is not almost everywhere zero.*

Denote $|x|$ the usual euclidean norm, and when A is any hermitian matrix, set

$$\begin{aligned} \|A\| &= \sup_{x \neq 0} \frac{|Ax|}{|x|} \\ &= \sup\{|\tau_j|, \text{ where } \tau_j \text{ is any eigenvalue of } A\} \\ &= \sup_{|x|=1} x^* A x. \end{aligned}$$

For any pair of hermitian matrices A, B we write $A \geqslant B$ whenever $A - B$ is hermitian nonnegative.

When $\varphi \geqslant \psi$, one has thus, by linearity, $T_n(\varphi) \geqslant T_n(\psi)$. For bounded φ, this implies

$$\|T_n(\varphi)\| \leqslant \sup_{\lambda \in \mathbb{T}} |\varphi(\lambda)| = \|\varphi\|_\infty \, .$$

Let X be a real-valued, second order stationary process. From (3) and (5) we get

(9) $$2\pi I_n(\varphi) = \frac{1}{n}\int_{\mathbb{T}} \left|\sum_{k=1}^{n} X_k e^{-ik\lambda}\right|^2 \varphi(\lambda)d\lambda$$

whence, in view of (8) and with the notation

$$X(n) = \begin{bmatrix} X_1 \\ \vdots \\ X_n \end{bmatrix}$$

(10) $$I_n(\varphi) = \frac{1}{n} X(n)^* T_n(\varphi) X(n).$$

For *odd functions* φ, i.e. such that $\varphi(-\lambda) = -\varphi(\lambda)$, we have $I_n(\varphi) = 0$ by (9). *For even functions* φ, $T_n(\varphi)$ *is real and symmetric.* Let us refine (10) in this case.

Lemma. *Let X be a centered stationary gaussian process, with spectral density f. Let φ: $\mathbb{T} \to \mathbb{R}$ be a (Lebesgue) integrable even function. Then for each n, one has*

$$I_n(\varphi) = \frac{1}{n}\sum_{k=1}^{n} \tau_k Y_k^2$$

where the τ_k are the eigenvalues of $T_n(2\pi f)^{1/2} T_n(\varphi) T_n(2\pi f)^{1/2}$ and the Y_k are independent random variables having the same distribution $N(0,1)$.

Proof. We may obviously restrict ourselves to the case when X is not identically zero. The covariance matrix of $X(n)$ is

$$\Gamma_n = E[X(n)X(n)^*] = T_n(2\pi f)$$

and is hence positive definite since $f \not\equiv 0$. The symmetric matrix $A = \Gamma_n^{1/2} T_n(\varphi)\Gamma_n^{1/2}$ may be written $A = Q^*DQ$ where D is a diagonal matrix and Q is orthogonal. Whence, with the definition

$$Y = Q\Gamma_n^{-1/2}X(n) = \begin{bmatrix} Y_1 \\ \vdots \\ Y_n \end{bmatrix}$$

$$I_n(\varphi) = \frac{1}{n} X(n)^* T_n(\varphi) X(n) = \frac{1}{n} Y^* D Y = \frac{1}{n}\sum_{k=1}^{n} \tau_k Y_k^2$$

and Y is clearly centered gaussian with covariance matrix equal to the unit matrix.

3. Asymptotic Normality and Periodogram

In the case where X is gaussian, its periodogram I_n has an asymptotically gaussian distribution, i.e. for every finite family of functions φ, the joint distribution of the $\sqrt{n}[I_n(\varphi) - I(\varphi)]$ is gaussian for n large. Moreover for φ_1 and φ_2 with disjoint supports, the corresponding random variables are asymptotically independent. This gives a heuristic explanation of the nonconvergence of $I_n(\lambda)$. If $I_n(\lambda)$ did converge toward $f(\lambda)$, the process $\sqrt{n}[I_n(\lambda) - f(\lambda)]$ "would converge" toward a totally chaotic gaussian process Z_λ, $\lambda \in \mathbb{T}$ where all the Z_λ would be independent. We shall see that such a process would "look like" the derivative of a brownian process, which happens not to have any derivative.

For every bounded function φ, the function

$$\tilde{\varphi}(\lambda) = \frac{1}{2}[\varphi(\lambda) + \varphi(-\lambda)]$$

is even and $I_n(\varphi) = I_n(\tilde{\varphi})$. We shall hence restrict ourselves to *the case where* φ *is even.*

For every integrable function g: $\mathbb{T} \to \mathbb{R}$ with Fourier coefficients $\hat{g}_k$, set

$$\alpha(g) = \sum_{k \in \mathbf{Z}} (1 + |k|)|\hat{g}_k|. \tag{11}$$

The norm $\alpha(g)$ is finite if g is smooth enough; for instance if g''' exists and is bounded, with g periodic of period 2π, one has $|\hat{g}_k| \leq c/(1 + |k|)^3$ and hence $\alpha(g) < \infty$. We shall impose, on the spectral density f of X, the restriction $\alpha(f) < \infty$.

For every bounded *vector valued* function φ: $\mathbb{T} \to \mathbb{R}^k$, we define as in (5) and (6) the k-dimensional *random vectors* $I_n(\varphi)$ and the *vector* $I(\varphi) \in \mathbb{R}^k$.

3.1. Theorem. *Let X be a centered gaussian process with covariances r_k verifying $\Sigma_k|k||r_k| < \infty$. Let f be the spectral density of X. Then for every bounded even Borel function φ: $\mathbb{T} \to \mathbb{R}^k$, one has*

$$\lim_{n\to\infty} E[I_n(\varphi)] = I(\varphi)$$

and the distribution of $\sqrt{n}[I_n(\varphi) - I(\varphi)]$ converges tightly, as $n \to \infty$, toward the gaussian distribution $N[0,\Gamma(\varphi)]$ where

$$\Gamma(\varphi) = 4\pi \int_{\mathbb{T}} \varphi\varphi^* f^2 d\lambda.$$

Moreover if φ is continuous, the sequence $I_n(\varphi)$ converges almost surely toward $I(\varphi)$.

Proof. Set $V_n = I_n(\varphi)$, $V = I(\varphi)$. To obtain the sought-for convergence in distribution, it is enough, by Levy's theorem, to prove that for all $u \in \mathbb{R}^k$

$$\lim_{n\to\infty} E[\exp(i\sqrt{n}\, u^*(V_n - V))] = \exp\left[-\frac{1}{2}u^*\Gamma(\varphi)u\right]$$

which may be written, denoting by ψ: $\mathbb{T} \to \mathbb{R}$ the function $\psi = u^*\varphi$,

$$\lim_{n\to\infty} E[\exp(i\sqrt{n}[I_n(\psi) - I(\psi)])] = \exp\left[-2\pi\int_{\mathbb{T}} \psi^2 f^2 d\lambda\right].$$

We thus only need to prove that

(12) $$\lim_{n\to\infty} \sqrt{n}\, [E(I_n(\psi)) - I(\psi)] = 0$$

(13) $$\lim_{n\to\infty} E\{\exp(i\sqrt{n}[I_n(\psi) - E(I_n(\psi))])\} = \exp\left[-2\pi\int_{\mathbb{T}} \psi^2 f^2 d\lambda\right]$$

Formulas (2), (5), and (6) and Parseval's identity yield

$$\begin{aligned} &2\pi(E[I_n(\psi)] - I(\psi)) \\ &= \sum_{|k|\leq n-1}\left(1 - \frac{|k|}{n}\right) r_k\hat{\psi}_k - \sum_k r_k\hat{\psi}_k \\ &= -\sum_{|k|\leq n-1} \frac{|k|}{n} r_k\hat{\psi}_k - \sum_{|k|\geq n} r_k\hat{\psi}_k . \end{aligned}$$

Since $|\hat{\psi}_k| \leq \|\psi\|_\infty$, the term $(-\Sigma_{|k|\leq n-1} \ldots)$ is bounded in modulus by

$$\frac{1}{n}\|\psi\|_\infty\left[\sum_{|k|\leq n-1} |k||r_k|\right];$$

the term $(-\Sigma_{|k|\geq n} \ldots)$ is bounded in modulus by

$$\sum_{|k|\geq n} \frac{|k|}{n} |r_k| \, |\hat{\psi}_k|$$

and hence by

$$\frac{1}{n} \|\psi\|_\infty \left[\sum_{|k|\geq n} |k| \, |r_k| \right].$$

We thus obtain

$$|E(I_n(\psi)) - I(\psi)| \leq \frac{1}{n} \|\psi\|_\infty \, \alpha(f)$$

which proves (12.).

By Lemma 2.2, we may write

$$I_n(\psi) = \frac{1}{n} \sum_{k=1}^{n} \tau_k Y_k^2$$

where $Y = (Y_1 \ldots Y_n)$ is standard gaussian, and the τ_k are the eigenvalues of

$$A(\psi) = [T_n(2\pi f)]^{1/2} T_n(\psi) [T_n(2\pi f)]^{1/2};$$

consequently

$$E[I_n(\psi)] = \frac{1}{n} \sum_{k=1}^{n} \tau_k = \frac{1}{n} \operatorname{tr} A(\psi).$$

Every r.v. W with distribution $N(0,1)$ verifies

$$E[\exp iv(W^2 - 1)] = \frac{1}{\sqrt{1-2iv}} \exp(-iv), \qquad \text{for } v \in \mathbb{R},$$

whence, the Y_j being independent,

$$\log E \exp i\sqrt{n}(I_n(\psi) - E[I_n(\psi)])$$

$$= -\sum_{k=1}^{n} \left[i \frac{\tau_k}{\sqrt{n}} + \frac{1}{2} \log \left[1 - 2i \frac{\tau_k}{\sqrt{n}} \right] \right].$$

The eigenvalues τ_k of $A(\psi)$ are bounded by $\|A(\psi)\|$ and hence by

$$\|[T_n(2\pi f)]^{1/2}\|^2 \|T_n(\psi)\| \leq 2\pi \|f\|_\infty \|\psi\|_\infty.$$

The Taylor expansion of order 2 for $\log(1 + z)$ yields then

$$\log E \exp i\sqrt{n}(I_n(\psi) - E[I_n(\psi)]) = -\frac{1}{n}\sum_{k=1}^{n} \tau_k^2 + R_n$$

with

$$|R_n| \leqslant (\text{constant}) \times \frac{1}{n^{3/2}} \sum_{k=1}^{n} |\tau_k|^3.$$

Since the $|\tau_k|$ are bounded, and since tr $A^2(\psi) = \Sigma_{k=1}^n \tau_k^2$, assertion (13) will be proved if we show that

$$(14) \qquad \lim_{n\to\infty} \frac{1}{n} \operatorname{tr} A^2(\psi) = 2\pi \int \psi^2 f^2 d\lambda.$$

Assume first that $\alpha(\psi)$ *is finite.* The matrix $A^2(\psi)$ has the same trace as $T_n(2\pi f)T_n(\psi)T_n(2\pi f)T_n(\psi)$ thanks to the elementary formula tr BC = tr CB. *For* $\alpha(f)$ and $\alpha(\psi)$ finite, we have, due to the results of Section 5 below,

$$\lim_{n\to\infty} \frac{1}{n} \operatorname{tr}\{[T_n(f)T_n(\psi)]^2 - T_n(f^2\psi^2)\} = 0;$$

hence

$$\frac{1}{n} \operatorname{tr} A^2(\psi) \quad \text{and} \quad \frac{1}{n} \operatorname{tr} T_n(4\pi^2 f^2 \psi^2)$$

have the same limit, which proves (14) since one always has (cf. (7))

$$\frac{1}{n} \operatorname{tr} T_n(g) \equiv \frac{1}{2\pi}\int_{\mathbb{T}} g(\lambda) d\lambda.$$

For real and symmetric matrices A, B such that $0 \leqslant B \leqslant A$ one has, using again tr A_1A_2 = tr A_2A_1,

$$\operatorname{tr}(A^2 - B^2) = \operatorname{tr}(A + B)(A - B)$$

$$= \operatorname{tr}(A + B)^{1/2}(A - B)(A + B)^{1/2} \geqslant 0.$$

For $0 \leqslant \psi_1 \leqslant \psi_2$ one has $0 \leqslant T_n(\psi_1) \leqslant T_n(\psi_2)$ whence $0 \leqslant A(\psi_1) \leqslant A(\psi_2)$ and thus tr $A^2(\psi_1) \leqslant$ *tr* $A^2(\psi_2)$. When ψ is a bounded Borel function, with lower bound $m > 0$, we can always find sequences φ_k, ψ_k with $\alpha(\varphi_k)$, $\alpha(\psi_k)$ finite, and $0 < \psi_k \leqslant \psi \leqslant \varphi_k$ such that

$$\alpha_k = \int(\psi_k^2(\lambda) - \varphi_k^2(\lambda))f(\lambda)d\lambda$$

converges toward 0 as $k \to \infty$. Indeed, since ψ is bounded, ψ is

(Lebesgue) almost everywhere the pointwise limit of an increasing sequence of positive continuous functions ψ_k and of a decreasing sequence of uniformly bounded continuous functions φ_k. The functions φ such that $\alpha(\varphi) < \infty$ being dense in the space of continuous functions (for the uniform norm), we may choose ψ_k, φ_k such that $\alpha(\psi_k)$ and $\alpha(\varphi_k)$ are finite. The almost everywhere pointwise convergence of $(\varphi_k - \psi_k)$ toward zero implies then the convergence of α_k toward zero. This yields lower and upper bounds for the limits $\underline{\lim}_{n\to\infty}$ and $\overline{\lim}_{n\to\infty}$ of $(1/n)\mathrm{tr}\, A^2(\psi)$, and proves (14) in this case.

Finally an elementary computation (by polarization) proves that if ψ, η, and $(\psi + \eta)$ verify (14) then $(\psi - \eta)$ also verifies (14), which proves (14) for all bounded Borel functions ψ, and yields the announced asymptotic normality for $\sqrt{n}[I_n(\varphi) - I(\varphi)]$.

The almost sure convergence of $I_n(\varphi)$ toward $I(\varphi)$ for continuous φ: $\mathbb{T} \to \mathbb{R}^k$ is a consequence of Theorem 2.2, since the hypothesis on the r_k implies the ergodicity of the gaussian process X as seen in Chapter 3, Theorem 4.4.

3.3. Corollary. *Let X be gaussian as in* 3.2. *Then if $J_1 \dots J_k$ are disjoint intervals in* $\mathbb{T}$, *the* r.v.

$$\int_{J_1} I_n(\lambda)d\lambda\,, \ \dots\,,\ \int_{J_k} I_n(\lambda)d\lambda$$

are asymptotically independent.

4. Asymptotic Normality of Empirical Estimators

4.1. Theorem. *Let X be a centered stationary gaussian process, with covariances r_k verifying $\Sigma_k |k||r_k| < \infty$. Then for each fixed k the distribution of the random vector*

$$\sqrt{n}[\hat{r}_0(n) - r_0,\ \hat{r}_1(n) - r_1,\ \dots,\ \hat{r}_k(n) - r_k]$$

converges tightly toward the gaussian distribution $N(0,\Gamma)$ where $\Gamma = [\Gamma_{st}]$, $0 \leqslant s,t \leqslant k$ and

$$\Gamma_{st} = 4\pi \int_{\mathbb{T}} \cos(s\lambda)\cos(t\lambda) f^2(\lambda)d\lambda$$

$$= \sum_{m\in\mathbb{Z}} (r_m r_{m+s+t} + r_m r_{m+s-t}).$$

Proof. Definition (2) shows that $\hat{r}_j(n) = I_n(\varphi_j)$ where $\varphi_j(\lambda) = \cos(j\lambda)$ which allows the use of 3.2. To get an explicit expression of Γ_{st}, one replaces $4\pi^2 f^2$ by its Fourier series

$$(2\pi f(\lambda))^2 = \left[\sum_j r_j e^{-ij\lambda}\right]^2 = \sum_j \left[\sum_m r_m r_{j-m}\right] e^{-ij\lambda}.$$

From this theorem, we are now going to deduce the asymptotic normality of all empirical estimators, since asymptotic normality is carried over by differentiable functions. First let us recall this classical result.

4.2. Theorem. *Let $u \in \mathbb{R}^k$ and let u_n be a sequence of random vectors with values in $\mathbb{R}^k$, such that the distribution of $\sqrt{n}(u_n - u)$ converges tightly, as $n \to \infty$, toward the gaussian distribution $N(0,\Gamma)$. Let f: $\mathbb{R}^k \to \mathbb{R}^\ell$ be a function, of class 2 in the neighborhood of u; set $v = f(u)$ and $v_n = f(u_n)$. Then, as $n \to \infty$, the distribution of $\sqrt{n}(v_n - v)$ converges tightly toward $N(0,\tilde{\Gamma})$ where $\tilde{\Gamma} = f'(u)\Gamma f'(u)^*$.*

Proof. Setting $f'(u) = A$, $y_n = \sqrt{n}(v_n - v)$, $x_n = \sqrt{n}(u_n - u)$, the Taylor expansion of f at the point u yields the relation

$$y_n = Ax_n + \frac{1}{\sqrt{n}} R_n \qquad \text{for } |x_n| \leqslant \rho$$

with a remainder $|R_n|$ bounded by $\rho^2 \sup_{|z| \leqslant \rho} \|f''(z)\|$. Since $P(|x_n| > \rho)$ converges toward zero, we immediately obtain, for any $w \in \mathbb{R}$,

$$\lim_{n\to\infty} E(e^{iw^*y_n}) = \lim_{n\to\infty} E(e^{iw^*Ax_n}) = \lim_{n\to\infty} \varphi_n(A^*w)$$

where φ_n is the characteristic function of x_n. By hypothesis we have

$$\lim_{n\to\infty} \varphi_n(z) = \exp\left[-\frac{1}{2} z^*\Gamma z\right]$$

for $z \in \mathbb{R}^k$, whence the conclusion by Levy's theorem.

4.3. Corollary. *Let X be a centered stationary gaussian process, with covariances verifying $\Sigma_k |k| r_k < \infty$. Then for every integer k, and every function f: $\mathbb{R}^{k+1} \to \mathbb{R}^\ell$ which is twice continuously differentiable in the neighborhood of $(r_0 \ldots r_k)$, the empirical estimator $\hat{v}_n = f(\hat{r}_0(n) \ldots \hat{r}_k(n))$ is a consistent estimator of $v = f(r_0 \ldots r_k)$, and the distribution of $\sqrt{n}(\hat{v}_n - v)$ converges tightly, as*

$n \to \infty$, toward the gaussian distribution $N(0, A\Gamma A^*)$ where $A = f'(r_0 ... r_k)$ and Γ is given by 4.1.

4.4. Corollary. *Let X be a gaussian process as in* 4.3. *Call $\rho_k = r_k / r_0$ the correlation of order k for X, and*

$$\hat{\rho}_k(n) = \frac{\hat{r}_k(n)}{\hat{r}_0(n)}$$

its empirical estimator. Then, as $n \to \infty$, the law of $\sqrt{n}[\hat{\rho}_1(n) - \rho_1, ..., \hat{\rho}_k(n) - \rho_k]$ converges toward $N(0,\Delta)$ where Δ is obtained as in 4.3. *Setting $\Delta = [\Delta_{st}]$ and*

$$\gamma_{st} = \sum_{m \in \mathbf{Z}} (\rho_m \rho_{m+s+t} + \rho_m \rho_{m+s-t})$$

we have

$$\Delta_{st} = \gamma_{st} - \gamma_{0s}\rho_t - \gamma_{0t}\rho_s + \gamma_{00}\rho_s\rho_t.$$

5. *The Toeplitz Asymptotic Homomorphism*

We are going to prove that for n large the map $\varphi \to (1/n)T_n(\varphi)$ is approximately an algebra homomorphism, as suggested by the formula

$$(15) \qquad (\varphi\psi)_k = \sum_j \hat{\varphi}_j \hat{\psi}_{k-j}, \qquad k \in \mathbb{Z},$$

true for arbitrary square integrable functions $\varphi, \psi: \mathbb{T} \to \mathbb{R}$, and obtained by multiplying the Fourier series of φ and ψ. The notations are those of 2.2.

From (15) one deduces easily that the norm $\alpha(\varphi)$ defined by (11) verifies

$$(16) \qquad \alpha(\varphi\psi) \leqslant \alpha(\varphi)\alpha(\psi)$$

so that the set of all φ such that $\alpha(\varphi) < \infty$ is an algebra $\mathcal{A}$. We introduce *three norms for square matrices*: if $A = (a_{ij})$ is a square matrix of order n with complex coefficients, we set

$$b(A) = \sum_{i,j} |a_{ij}| \qquad \textbf{(block norm)}$$

(17) $$\ell(A) = \sup_i \Big[\sum_j |a_{ij}|\Big] \qquad \textbf{(line norm)}$$

$$c(A) = \sup_j \Big[\sum_i |a_{ij}|\Big] \qquad \textbf{(column norm).}$$

These definitions imply the elementary inequalities

(18)
$$b(AB) \leqslant b(A)\ell(B) \leqslant b(A)b(B)$$

$$b(AB) \leqslant c(A)b(B) \leqslant b(A)b(B)$$

$$\ell(AB) \leqslant \ell(A)\ell(B) \quad \text{and} \quad c(AB) \leqslant c(A)c(B)$$

$$\operatorname{tr}(A) \leqslant b(A).$$

5.1. Proposition. *For every $k \geqslant 1$, $n \geqslant 1$, and $g_1 \ldots g_k \in \mathcal{A}$, one has the block norm bound*

$$b[T_n(g_1)\ldots T_n(g_k) - T_n(g_1 \ldots g_k)] \leqslant (k-1)\alpha(g_1)\ldots\alpha(g_k).$$

Proof. Take first $k = 2$. Let $\varphi, \psi \in \mathcal{A}$. By (15) one has, with the notation $\eta = \varphi\psi$,

$$\sum_{1 \leqslant i,k \leqslant n} \Big| \hat{\eta}_{i-k} - \sum_{1 \leqslant j \leqslant n} \hat{\varphi}_{i-j}\hat{\psi}_{j-k} \Big|$$

$$= \sum_{1 \leqslant i,k \leqslant n} \Big| \sum_{j \leqslant 0} \hat{\varphi}_{i-j}\hat{\psi}_{j-k} + \sum_{j \geqslant n+1} \hat{\varphi}_{i-j}\hat{\psi}_{j-k} \Big|.$$

Since $|\hat{\varphi}_\ell| = |\hat{\varphi}_{-\ell}|$ one has

$$\sum_{1 \leqslant i,k \leqslant n} \Big| \sum_{j \leqslant 0} \hat{\varphi}_{i-j}\hat{\psi}_{j-k} \Big| \leqslant \sum_{1 \leqslant i,k \leqslant n} \sum_{j \geqslant 0} |\hat{\varphi}_{i+j}|\, |\hat{\psi}_{k+j}|$$
$$= \sum_{j \geqslant 0} \rho_j(\varphi)\rho_j(\psi)$$

where

$$\rho_j(\varphi) = \sum_{i \geqslant 1} |\hat{\varphi}_{i+j}|.$$

A similar computation gives the same bound for the second sum above whence

(19) $$b[T_n(\varphi\psi) - T_n(\varphi)T_n(\psi)] \leqslant 2 \sum_{j \geqslant 0} \rho_j(\varphi)\rho_j(\psi).$$

The relations

$$\rho_j(\varphi) \leqslant \sum_{j \geqslant 1} |\hat{\varphi}_k| \quad \text{and} \quad \sum_{j \geqslant 0} \rho_j(\psi) = \sum_{k \geqslant 1} k|\hat{\psi}_k|$$

yield, for the right-hand side of (19), the bound

$$\left(\sum_{k \geqslant 1} |\hat{\varphi}_k|\right)\left(\sum_{k \geqslant 1} k|\hat{\psi}_k|\right) + \left(\sum_{k \geqslant 1} k|\hat{\varphi}_k|\right)\left(\sum_{k \geqslant 1} |\hat{\psi}_k|\right) \leqslant \alpha(\varphi)\alpha(\psi).$$

This proves the announced result when $k = 2$. We then point out that when n varies, the $T_n(\varphi)$ remain bounded in line norm and column norm since

(20) $$c[T_n(\varphi)] = \ell[T_n(\varphi)] = \sum_k |\hat{\varphi}_k| \leqslant \alpha(\varphi)$$

for all $n \geqslant 1$, which allows a proof by recursion on k, thanks to $b(AB) \leqslant b(A)\ell(B)$.

5.2. Corollary (Szegö). *Consider* φ: $\mathbb{T} \to \mathbb{R}$ *with* $\alpha(\varphi) < \infty$, *which implies the existence of* $[a,b]$ *such that* $a \leqslant \varphi \leqslant b$. *Then the eigenvalues* $\tau_1 \ldots \tau_n$ *of* $T_n(\varphi)$ *belong to* $[a,b]$, *and for every continuous function* g: $[a,b] \to \mathbb{R}$ *one has*

$$\lim_{n \to \infty} \frac{1}{n}[g(\tau_1) + \ldots + g(\tau_n)] = \frac{1}{2\pi}\int_{\mathbb{T}} g \circ \varphi(\lambda)d\lambda.$$

Proof. This is equivalent to the tight convergence of the empirical distribution

$$\nu_n = \frac{1}{n}(\delta_{\tau_1} + \ldots + \delta_{\tau_n})$$

toward the measure image of $(1/2\pi\eta)$ by φ, where η is Lebesgue measure on . Classically it suffices to check (cf. [36], p. 185) that the moments of ν_n converge toward those of ν, that is to deal with the case where $g(x) = x^p$. One has then

$$\int_{\mathbb{R}} g\, d\nu_n = \frac{1}{n}(\tau_1^p + \ldots + \tau_n^p) = \frac{1}{n}\operatorname{tr}\{T_n(\varphi)^p\}.$$

By 5.1, and the bound $\operatorname{tr}(A) \leqslant b(A)$ we have

$$\lim_{n \to \infty} \frac{1}{n} \operatorname{tr}\{[T_n(\varphi)]^p - T_n(\varphi^p)\} = 0$$

whence the result since

$$\frac{1}{n} \operatorname{tr} T_n(\varphi^p) \equiv \frac{1}{2\pi}\int_{\mathbb{T}} \varphi^p(\lambda)d\lambda = \int_{\mathbb{R}} g\, d\nu.$$

5.3. Corollary. *Let X be a centered stationary process (not necessarily gaussian), with strictly positve spectral density f such that $\alpha(f) < \infty$. Then the one-step prediction error σ^2 is given by*

$$\log \sigma^2 = \lim_{n\to\infty} \frac{1}{n} \log \det T_n(2\pi f) = \frac{1}{2\pi} \int_{\mathrm{T}} \log 2\pi f(\lambda) d\lambda.$$

Proof. We have seen (Chapter 4, Section 4.3) that

$$\log \sigma^2 = \lim_{n\to\infty} \frac{1}{n} \log \det \Gamma_n$$

where Γ_n is the covariance of

$$X(n) = \begin{bmatrix} X_1 \\ \vdots \\ X_n \end{bmatrix}.$$

Since $\Gamma_n = T_n(2\pi f)$, we obtain the following relation, where $\tau_1 \ldots \tau_n$ are the eigenvalues of Γ_n,

$$\log \sigma^2 = \lim_{n\to\infty} \frac{1}{n} [\log \tau_1 + \ldots + \log \tau_n].$$

But the hypothesis "$\alpha(f)$ finite" implies the continuity of f, which in turn implies, since $f > 0$, the existence of a, b such that $0 < a \leqslant f \leqslant b$, and the result is a consequence of 5.2.

Bibliographical Hints

This chapter focuses on the convergence of empirical spectral measures $I_n(\lambda)d\lambda$ toward $f(\lambda)d\lambda$. The regularity condition $\alpha(f) < \infty$ imposed on f simplifies the computations on Toeplitz forms and leads to the Szego isomorphism as well as the value of

$$\lim_{n\to\infty} \frac{1}{n} \log \det \Gamma_n .$$

This type of idea has been introduced in statistics by *Grenander* and *Szegö*. The condition $\alpha(f) < \infty$ may be weakened by the use of more refined Banach algebra techniques, see *Widom.* The nonconvergence of $I_n(\lambda)$ toward $f(\lambda)$ is analyzed in details by *Ohlsen* where the general case is reduced to the white noise case, since $I_n^X(\lambda)$ is well-approximated by $f(\lambda)I_n^W(\lambda)$ where W is the innovation of X. Nevertheless, this approach is not the most efficient to study the limit of $I_n(\lambda)d\lambda$. The study of the variance of $\sqrt{n}[I_n(\varphi) - I(\varphi)]$ is easier for f continuous, the study of the bias $E[I_n(\varphi)] - I(\varphi)$ for f with bounded variation. A very clear presentation is given by *Ibraguimov*, and is followed by *Dacunha-Castelle* and *Duflo* (these techniques do not lead to the Szegö isomorphism which necessitates an underlying Banach algebra).

As indicated in the introduction, the estimation of $f(\lambda)$ is not studied in this book. Instead of $I_n(\lambda)$ whose variance does not converge to zero, or of $\int I_n(\lambda')\varphi(\lambda - \lambda')d\lambda'$, whose bias does not converge toward zero, one uses the so-called window estimates $\int I_n(\lambda')\varphi_n(\lambda - \lambda')d\lambda'$ with $\int \varphi_n(\lambda)d\lambda = 1$ and

$$\lim_{n\to\infty} \int_{|\lambda|\geq a>0} \varphi_n(\lambda)d\lambda = 0 \qquad \text{for all } a > 0.$$

There is a large literature on the subject: *T. W. Anderson*, *Priestly* and *Koopmans*. *Hannan's* Chapter V is quite detailed, and studies also estimators of f often used in physics, which rely on Fourier coefficients estimation by the so-called fast Fourier transform.

Chapter XI
EMPIRICAL ESTIMATION OF THE PARAMETERS FOR ARMA PROCESSES WITH RATIONAL SPECTRUM

1. Empirical Estimation and Efficient Estimation

Let X be a centered stationary process, with rational spectrum. We seek to estimate, starting from observations $X_1 ... X_N$, the natural parameters identifying the covariance structure of X. Write *the canonical ARMA relation* linking X and its innovation W

$$(1) \qquad \sum_{k=0}^{p} a_k X_{n-k} = \sum_{\ell=0}^{q} b_\ell W_{n-\ell}$$

where

$$P(z) = \sum_{k=0}^{p} a_k z^k \quad \text{and} \quad Q(z) = \sum_{\ell=0}^{q} b_\ell z^\ell$$

are coprime, the roots of P have moduli > 1, the roots of Q have moduli $\geqslant 1$ and $a_0 = b_0 = 1$, $a_p \neq 0$, $b_q \neq 0$. Call σ^2 the variance of W.

The point is to select p, q and then to estimate $a_1 ... a_p$, $b_1 ... b_q$, σ^2. A first approach is to express these parameters in terms of the covariances $r_k = E(X_n X_{n+k})$ of X, and then as in Chapter 10, Section 1, to replace in these "formulas" the r_k by their usual estimators $\hat{r}_k(n)$. For gaussian processes X, we may expect (cf. Chapter 10) these estimators to be consistent and asymptotically normal, but in general *they will not be efficient, i.e. they will not have minimal asymptotic variance.* We shall use these empirical estimators as **preliminary** estimators, which will constitute the first step in the computation of efficient estimators supplied by the maximum likelihood method (cf. Chapter 13). We shall come back, in Section 11 of this chapter, to the notion of empirical estimator.

2. Computation of the a_k and Yule-Walker Equations

Let X have rational spectrum and minimal type (p,q) with $p \geqslant 1$; we keep the notations of Section 1. Multiplying (1) by X_{n-m} we obtain

$$\sum_{k=0}^{p} a_k E(X_{n-k} X_{n-m}) = \sum_{\ell=0}^{q} b_\ell E(W_{n-\ell} X_{n-m}).$$

Since W is the innovation of X, one has $E(W_{n-\ell} X_{n-m}) = 0$ for $m \geqslant \ell + 1$. The covariances r_m hence verify the recursive relation

(2) $$r_m + a_1 r_{m-1} + \dots + a_p r_{m-p} = 0 \qquad \text{for } m \geqslant q+1.$$

For $q + 1 \leqslant m \leqslant q + p$ we thus obtain a system of p linear equations in $a_1 \ldots a_p$ which may be written (Yule-Walker equations)

$$\begin{bmatrix} r_q & r_{q-1} & \cdots & r_{q-p+1} \\ r_{q+1} & r_q & \cdots & r_{q-p+2} \\ \vdots & \vdots & & \vdots \\ r_{q+p-1} & r_{q+p-2} & \cdots & r_q \end{bmatrix} \begin{bmatrix} a_1 \\ \vdots \\ a_p \end{bmatrix} = - \begin{bmatrix} r_{q+1} \\ \vdots \\ r_{q+p} \end{bmatrix}$$

which we shall write

(4) $$R(p,q)a = -r(p,q).$$

Since X is of minimal type (p,q) we shall see in 5.3 and 5.4 that $R(p,q)$ is necessarily invertible. Whence the explicit expression

(5) $$a = g(r_{q-p+1}, r_{q-p+2}, \dots, r_{q+p}) = R(p,q)^{-1} r(p,q)$$

where the coordinates of $g: \mathbb{R}^{2p} \to \mathbb{R}^p$ are rational fractions.

3. Computation of the b_ℓ and of σ^2

3.1. A System of Quadratic Equations

Assume the a_k to be known, or computed by (5). Set $Y_n = \Sigma_{k=0}^{p} a_k X_{n-k}$, which yields directly the covariances s_j of Y

$$(6)\qquad \begin{aligned} s_j = E(Y_n Y_{n+j}) &= \sum_{0 \leqslant k,\ell \leqslant p} a_k a_\ell E(X_{n-k} X_{n+j-\ell}) \\ &= \sum_{0 \leqslant k,\ell \leqslant p} a_k a_\ell r_{j+k-\ell}. \end{aligned}$$

On the other hand, by (1) we also have

$$s_j = E\left[\left[\sum_{\ell=0}^{q} b_\ell W_{n-\ell}\right]\left[\sum_{u=0}^{q} b_u W_{n+j-u}\right]\right].$$

Since W is a white noise with variance σ^2, we get

$$(7)\qquad \sigma^2 \sum_{u=j}^{q} b_u b_{u-j} = s_j, \quad 0 \leqslant j \leqslant q$$

and the s_j are zero for $j \geqslant q + 1$.

The point is *to solve system* (7) *where the s_j are known, and σ^2, $b_1 \dots b_q$ are unknown* (recall that $b = 1$). Setting $x_0 = \sigma$ and $x_j = \sigma b_j$ for $1 \leqslant j \leqslant q$ this system becomes

$$(8)\qquad \sum_{u=j}^{q} x_u x_{u-j} = s_j \qquad 0 \leqslant j \leqslant q.$$

For $q \geqslant 2$ such a system cannot be solved by explicit formulas, and has in general a finite number of distinct solutions. Note moreover that *the sought-for solution is here such that the polynomial*

$$S(z) = \sum_{\ell=0}^{q} x_\ell z^\ell$$

has all its roots of modulus $\geqslant 1$. Indeed we have $S = \sigma Q$.

3.2. The Solutions of System (7)

Write the Fourier series of the spectral density f^Y of Y,

$$f^Y(\lambda) = \frac{1}{2\pi} \sum_k s_k e^{-ik\lambda} = \frac{1}{2\pi} \sum_{|k| \leqslant q} s_k e^{-ik\lambda}$$

since we have seen that the s_k are zero for $|k| \geqslant q + 1$. But

$Y = A_Q W$ forces

$$f^Y(\lambda) = \frac{\sigma^2}{2\pi} |Q(e^{-i\lambda})|^2$$

and, finally, setting $z = e^{-i\lambda}$

$$(9) \qquad |\sigma Q(z)|^2 = \sum_{|k| \leq q} s_k z^k \qquad \text{for } |z| = 1.$$

Expanding explicitly the left-hand side of (9) we conclude that *system* (7) *and relation* (9) *are equivalent.*

With the preceding change of variables, we see that *solving system* (8) *for* $x \in \mathbb{R}^{q+1}$ *is equivalent to finding the polynomials* $S(z) = \Sigma_{\ell=0}^{q} x_\ell z^\ell$ *which verify the identity*

$$(10) \qquad S(z)S(1/z) = s_0 + \sum_{1 \leq k \leq q} s_k \left[z^k + \frac{1}{z^k} \right] \qquad \text{for all } z \in \mathbb{C}.$$

Indeed for $|z| = 1$, (10) and (8) are clearly equivalent, and two rational fractions in z which coincide for $|z| = 1$ are identical.

By induction, it is easily seen that $z^k + 1/z^k$ is a polynomial of degree k in $(z + (1/z))$ and hence the right-hand side of (10) may be written $U(z + (1/z))$ where U is a polynomial of degree q whose (real-valued) coefficients are universal linear combinations of the s_k, $0 \leq k \leq q$.

Factorization (9) is precisely the Fejer-Riesz representation of f^Y. *Thus the solutions of system* (7) [*or* (8)] *are in one-to-one correspondence with the Fejer-Riesz representations of the type* $f^Y(\lambda) = |S(e^{-i\lambda})|^2$, *where S has real coefficients.* Lemma 4.5 of Chapter 8 then shows that *system* (7) [*or* (8)] *has a finite number of solutions. If* $S(z) = c(z - \tau_1)...(z - \tau_q)$ *is an arbitrary solution, then all the other solutions are obtained by replacing one or several factors* $(z - \tau_j)$ *by* $\pm|\tau_j|(z - (1/\bar{\tau}_j))$, taking care of course of keeping the coefficients of *S real valued.* But up to a choice of sign, *there is a single solution S such that* $S(z)$ *has all its roots of modulus* ≥ 1, and this is precisely the solution we wanted here.

3.3. Numerical Solution of System (7)

We prefer to work on (8). Box-Jenkins advocate the resolution of (8) by Newton's method, which boils down to considering the sequence $v_{n+1} = v_n - g'(v_n)^{-1} g(v_n)$ when one wants to solve the vector equation $g(v) = 0$. Here of course we set $v = (v_0 ... v_q)$, $g = (g_0 ... g_q)$ where

$$g_j(x) = -s_j + \sum_{u=j}^{q} x_u x_{u-j} .$$

When v_n converges to v_∞ one always has $g(v_\infty) = 0$, and the convergence is often realized in practice ([53]). The real problem is the fact that this only provides *one* solution of (8) and *this particular solution is not necessarily the good one*, i.e. the associated polynomial S may well have roots of modulus < 1.

One may start the iteration all over again, with different initial points, but one still has to discover (at random!) the "good" initial points. This becomes so time consuming in practice that several algorithms actually in use on microcomputers prefer to avoid this computation completely and replace it by an initial random guess of the $b_1 ... b_j$.

Let us point out that as soon as an arbitrary solution S_0 of (8) has been obtained, one may in principle construct the good solution σQ by the method described in 3.2, after having factorized $S_0(z)$ in $\mathbb{C}[z]$. But *this demands a new numerical computation* (without explicit formulas), *namely the computation of all the roots of* S_0.

Since *it is unavoidable, except in particular cases, to compute the roots of a polynomial of degree* q, we propose the following more logical method. Write (10) in the form

$$S(z) S\left(\frac{1}{z}\right) = U\left(z + \left(\frac{1}{z}\right)\right)$$

where the coefficients of U are very easy to obtain explicitly starting with the known s_j. Solve in $\mathbb{C}$ the equation (of degree q) $U(\xi) = 0$. For $q \leqslant 3$ this is done explicitly, for $q \geqslant 4$ several algorithms are available; one may for instance use a gradient method (cf. [53]) to minimize $|U(\xi)|^2$ which supplies a first root, and then allows the reduction to degree $(q-1)$ etc.

Once the distinct roots $\xi_1 ... \xi_s$ of $U(\xi) = 0$ are obtained, with their multiplicities $m_1 ... m_s$, we may solve in z he elementary second-degree equations

$$z + \frac{1}{z} = \xi_j \qquad 1 \leqslant j \leqslant s$$

whose roots are of the form η_j and $1/\eta_j$ with $|\eta_j| \geqslant 1$.

In the generic case one has in fact $|\eta_j| > 1$ and the good solution $S = \sigma Q$ becomes

$$S(z) = c \prod_{1 \leq j \leq s} (z - \eta_j)^{m_j}$$

where c is computed by $S(i) = U(0)$ where $i^2 = -1$.

In general we may have $|\eta_j| = 1$, which clearly forces m_j to be even, and the good solution $S = \sigma Q$ becomes

$$S(z) = c \prod_{|\eta_j|>1} (z - \eta_j)^{m_j} \prod_{|\eta_j|=1} [(z - \eta_j)(z - \bar{\eta}_j)]^{1/m_j} .$$

We shall propose further down in Section 8 an iterative method of computation for the b_ℓ *which has the advantage of being explicit, and to converge surely toward the good solution, while avoiding the resolution of algebraic equations.*

4. Empirical Estimation of the Parameters When *p,q* are Known

For known p,q we have obtained an explicit expression of $a = (a_1 ... a_p)$ as *an analytic function* $g(r)$ where $r = (r_0 ... r_{q+p})$; this is a consequence of (5) since the r_j, $q - p + 1 \leq j \leq q + p$ are equal to coordinates of r when $p \geq 1$, due to the relation $r_k = r_{-k}$.

In Section 3 we obtained the expression (6) of the s_j as polynomials in a and r; the coefficients of $U(z)$, being linear combinations of the s_j, must then be analytic functions of r since $a = g(r)$.

Assume that *the roots* η_j *of* Q *are of modulus* > 1 (instead of ≥ 1 which is the general case) *and are distinct.* Then the roots

$$\xi_j = \eta_j + \frac{1}{\eta_j}$$

of U are obviously distinct. But the function which associates the roots of a polynomial to its coefficients is analytic in the neighborhood of any polynomial having distinct roots (implicit functions theorem). Consequently the roots η_j, $1 \leq j \leq q$ of $S(z)$ obtained in Section 3.3 thanks to the relation $S(z)S(1/z) = U(z)$ are analytic functions of the coefficients of U in the neighborhood of the coefficients of $U_0(z) = \sigma Q(z) \sigma Q(1/z)$, which are themselves analytic functions of r. The transition from the roots of $S(z)$ to its coefficients $x_0 ... x_q$ is polynomial, and the transition from $x_0 ... x_q$ to $b_1 ... b_q$ and σ is achieved by rational fractions. Whence the result.

Lemma. *When the roots of Q are distinct and of modulus* > 1, *there is a function* h: $\mathbb{R}^{p+q+1} \to \mathbb{R}^{p+q+1}$, *analytic in the neighborhood of* $r = (r_0, ..., r_{q+p})$ *such that the vector of parameters* $\theta = (a_1...a_p, b_1...b_q, \sigma)$ *verifies* $\theta = h(r)$.

We then define an **empirical estimator** $\hat{\theta}_n$ of θ by $\hat{\theta}_n = h[\hat{r}(n)]$ with $\hat{r}(n) = [\hat{r}_0(n)...\hat{r}_{q+p}(n)]$ which amounts to the replacement of r_j by $\hat{r}_j(n)$ in the computations of Sections 2 and 3. The results of Chapter 10, Theorem 4.1 and 4.3 grant the asymptotic normality of $\hat{\theta}_n$.

4.1. Theorem. *Let X be a stationary gaussian process, with rational spectrum of minimal type* (p,q). *Assume the canonical* ARMA *relation* $A_P X = A_Q W$ *linking X and its innovation W is generic*, i.e. *that Q has all its roots distinct and of modulus* > 1. *Then the empirical estimator* $\hat{\theta}_n$ *of* $\theta = (a_1...a_p, b_1...b_q, \sigma)$ *converges and the distribution of* $\sqrt{n}(\hat{\theta}_n - \theta)$ *converges tightly, as* $n \to \infty$, *to the gaussian distribution* $N(0, A\Gamma A^*)$ *where* $A = h'(r)$, *with h as above, and where Γ is the asymptotic covariance matrix of* $\sqrt{n}[\hat{r}(n) - r]$ *given by Theorem* 4.1, *Chapter* 10.

4.2. Remark. *In practice, it is essential, when one computes* $\hat{\theta}_n$ *as in Sections* 2 *and* 3, *to make sure that the estimated polynomials* $\hat{P}(z)$ *and* $\hat{Q}(z)$ *have all their roots of modulus* > 1. Under the generic hypothesis 4.1, this will always be true "for n large enough" which can only be a moral satisfaction for n given and fixed. For $\hat{Q}$ the method of computation of the $\hat{b}_\ell$ which we have suggested includes automatically such a verification. For $\hat{P}$ the computation of $\hat{a}$ by resolution of the Yule-Walker equations with r replaced by $\hat{r}$ does not grant anything on the roots of $\hat{P}$, for n given and finite, and the verification is necessary, unless the $\hat{r}$ happen to be the exact correlations of some empirical process.

If the $\hat{P}$ just *computed* does not have all its roots of modulus > 1, one may either replace $\hat{P}$ by the closest polynomial $\tilde{P}$ having all its roots of modulus > 1.05, or if the distance between $\hat{P}$ and $\tilde{P}$ is too large, start the computations again with $p' > p$.

A very simple way of testing if $P(z) = \Sigma_{k=0}^{p} a_k z^k$ has its roots of modulus > 1 is to compute by induction generic solutions f of $\Sigma_{k=0}^{p} a_k f(n-k) = 0$, $n \geq 0$ with arbitrary generic starting point $f(0)...f(p-1)$, since *one must then have* (cf. Chapter 9, Section 1.1)

$$\lim_{n\to+\infty} f(n) = 0$$

with exponential speed. Moreover if this is not the case (cf. Chapter 9, Section 1.1), *one has then for f generic*

$$\lim_{n\to+\infty} \frac{1}{n} \log|f(n)| = -\log \rho$$

where ρ is the minimum of the moduli of the roots of P, which is useful for the resolution of $P(z) = 0$.

5. Characterization of p and q

Recall that a process X with rational spectrum is of minimal type (p,q) if in the canonical ARMA relation of X, the polynomials P, Q are of degrees p and q. By Lemma 4.5 and Theorem 2.3, Chapter 8, for every ARMA(p',q') relation between X and an arbitrary white noise W' one has then $p' \geq p$ and $q' \geq q$.

5.1. Theorem. The Case of Moving Averages MA(q). *A process X having rational spectrum is of minimal type* $(0,q)$ *if and only if the covariances* r_m *of X are zero for* $|m| \geq q + 1$, *with* $r_q \neq 0$.

Proof. Let X be of minimal type (p,q). Assume the r_m to be zero for $|m| \geq t$ with $r_t \neq 0$. The spectral density f of X may be written, letting $z = e^{-i\lambda}$

$$f(\lambda) = \sum_{|m|\leq t} r_m z^m = V\left[z + \frac{1}{z}\right] = F(z), \qquad \lambda \in \mathbb{T}$$

where V is (cf. Section 3) a polynomial with real coefficients, of degree t. In particular the rational fraction f has the unique pole $z = 0$. The ARMA canonical relation (1) yields

$$f(z) = \frac{\sigma^2}{2\pi}\left|\frac{Q}{P}(z)\right| = \frac{\sigma^2}{2\pi}\,\frac{Q(z)Q(1/z)}{P(z)P(1/z)} \qquad \text{for } |z| = 1$$

which implies

$$F(z) \equiv \frac{\sigma^2}{2\pi}\,\frac{Q(z)Q(1/z)}{P(z)P(1/z)}$$

for all $z \in \mathbb{C}$. The poles of F are then the roots of P, their inverses, and possibly $z = 0$. Consequently P is constant and

$p = 0$. A computation sketched in Section 3 then yields

$$F(z) = cQ(z)Q\left(\frac{1}{z}\right) = \sum_{|k|\leqslant q} u_k z^k, \qquad z \in \mathbb{C}$$

with $u_k = u_{-k}$ and $u_q \neq 0$, which shows that $t = \text{degree}(V) = q$.
The converse statement is obvious.

5.2. Recursive Sequences

Consider $p \geqslant 0$, $q \geqslant 0$. We shall say that *a numerical sequence* r_m, $m \in \mathbb{Z}$ *satisfies a* (p,q) *recursive identity* if there are real numbers $\alpha_0 \ldots \alpha_p$ with $\alpha_0 = 1$, $\alpha_p \neq 0$ such that

$$\alpha_0 r_m + \alpha_1 r_{m-1} + \ldots + \alpha_p r_{m-p} = 0 \quad \text{for} \quad m \geqslant q + 1.$$

We shall say that *the sequence* r_m, $m \in \mathbb{Z}$ satisfies a **minimal** (p,q) **recursive** identity if the r_m satisfy the recursive identity just written and if for every (p',q') recursive identity satisfied by the r_m, one has $p' \geqslant p$ and $q' \geqslant q$.

For $s \geqslant 1$, $t \geqslant 0$ *we define the square matrices* $R(s,t)$, *of order* s, *by* $R(1,t) = [r_t]$ and

$$\text{(11)} \qquad R(s,t) = \begin{bmatrix} r_t & r_{t-1} & \cdots & r_{t-s+1} \\ r_{t+1} & r_t & \cdots & r_{t-s+2} \\ \vdots & \vdots & & \vdots \\ r_{t+s-1} & r_{t+s-2} & \cdots & r_t \end{bmatrix}.$$

We introduce also *their determinants*

$$\text{(12)} \qquad \rho(s,t) = \det R(s,t).$$

5.3. Lemma. (after Grantmacher [22], and Beghim-Gourieroux-Montfort [7]). *For every numerical sequence* r_m, $m \in \mathbb{Z}$ *the four following properties are equivalent:*

(i) *the sequence* r_m *satisfies a minimal* (p,q) *recursive identity*

(ii) *one has*
$$\rho(s,t) = 0 \quad \text{for} \quad s \geqslant p + 1,\ t \geqslant q + 1$$
$$\rho(p,t) \neq 0 \quad \text{for} \quad t \geqslant q$$
$$\rho(s,q) \neq 0 \quad \text{for} \quad s \geqslant p.$$

(iii) *one has* $\rho(p+1, t) = 0$ *for* $t \geq q + 1$
$\rho(p+1, q) \neq 0$ *for* $\rho(p,q+1) \neq 0$.

(iv) *one has* $\rho(s, q+1) = 0$ *for* $s \geq p + 1$
$\rho(p+1, q) \neq 0$ $\rho(q, p+1) \neq 0$.

Proof. See Section 10 below.

5.4. Proposition (after [7]). *Let X be a process with rational spectrum. For X to be of minimal (p,q) type, it is necessary and sufficient that the covariances r_m satisfy a minimal (p,q) recursive identity; this recursive identity is then given by* (2).

Proof. For $p = 0$ the point has been proved in 5.1. Assume then that $p \geq 1$ and that X is of minimal (p,q) type. Let $\alpha_0 \dots \alpha_{p'}$ be the coefficients of an arbitrary (p',q') recursive identity satisfied by the r_m. Define $Y_n = X_n + \alpha_1 X_{n-1} + \dots + \alpha_{p'} X_{n-p'}$. The covariances s_m of Y then become

$$s_m = \sum_{0 \leq i,j \leq p'} \alpha_i \alpha_j r_{m+i-j} = \sum_{0 \leq i \leq p'} \alpha_i \left[\sum_{0 \leq j \leq p'} \alpha_j r_{m+i-j} \right].$$

The identity satisfied by the r_m shows that $s_m = 0$ for $|m| \geq q' + 1$. By 5.1, Y is then of $(0,q'')$ type with $q'' < q'$. Thus Y and its innovation $\tilde{W}$ are linked by the MA relation $Y = A_{\tilde{Q}} \tilde{W}$ with degree $\tilde{Q} = q''$, and hence X is linked to $\tilde{W}$ by an ARMA(p',q'') relation, which yields $p' \geq p$ and $q'' \geq q$.

Hence for every (p',q') recursive identity satisfied by the r_m, one has $p' \geq p$, $q' \geq q$. On the other hand we have seen that the r_m satisfy a (p,q) recursive identity given by (2); this identity is hence a minimal (p,q) recursive identity.

To prove the converse, it is enough to notice that if a numerical sequence satisfies two minimal recursive relations of degrees (p_0,q_0) and (p,q), one has trivially $p_0 = p$, $q_0 = q$.

5.5. The Autoregressive Case

Assume we want to fit a minimal AR(s) model with coefficients $\alpha_0 = 1$, $\alpha_1 \dots \alpha_s$ to a process with rational spectrum, for which one only knows the covariances, with no information on its minimal type (p,q).

For each integer $s \geq 1$ we may a priori write the Yule-Walker equations (4) of order $(s,0)$

(13) $R(s,0)\,\alpha = -r(s,0)$

where the unknown α has coordinates $\alpha_1 \dots \alpha_s$. The matrix $R(s,0)$ coincides by definition with the covariance matrix of $X_1 \dots X_s$, and hence with the Toeplitz matrix $T_s(2\pi f)$ whence f is the spectral density of X. Hence $R(s,0)$ is invertible if f is not identically zero (cf. Chapter 10, Section 2.3), that is if X is not identically zero. Write then $\alpha(s) = -R(s,0)^{-1} r(s,0)$ and let $\Phi_s = \alpha_s(s)$ be the last coordinate of $\alpha(s)$. One calls Φ_s the **partial autocorrelation of order** s of X since (cf. Theorem 4, Section 4.1) *the r.v.* $\Sigma_{k=1}^{s} \alpha_k X_{n-k}$ *is the linear regression of* X_n *on* $X_{n-1} X_{n-2} \dots X_{n-s}$.

5.6. Proposition. *Let* X *be a process with rational spectrum. For* X *to be autoregressive and of minimal type* $(p,0)$ *it is necessary and sufficient that its partial autocorrelations* Φ_s *be zero for* $s \geqslant p+1$ *with* $\Phi_p \neq 0$.

Proof. Cramer's formulas for the resolution of the linear system (13), and definition (11) yield

$$\Phi_s = (-1)^s \frac{\det R(s,1)}{\det R(s,0)} .$$

Since $\rho(s,0) = \det R(s,0) \neq 0$, the hypothesis $\{\Phi_s = 0 \text{ for } s \geqslant p+1$ and $\Phi_p \neq 0\}$ is equivalent to $\{\rho(s,1) = 0 \text{ for } s \geqslant p+1 \text{ and } \rho(p,1) \neq 0\}$. In view of $\rho(p+1, 0) \neq 0$ and of the characterization 5.3(iv) this is equivalent to stating that the covariances r_m verify a minimal $(p,0)$-recursion, whence the result by 5.4.

6. Empirical Estimation of *d* for an ARIMA(*p,d,q*) Model

6.1. Empirical Estimation of Φ_s and $\rho(s,t)$

The empirical estimation of (p,d,q) for an ARIMA(p,d,q) process will be based on the set of indices s,t for which the Φ_s, $\rho(s,t)$, r_t are zero. We shall thus need simultaneous confidence intervals around the empirical estimators of these parameters.

The empirical estimator $\hat{\rho}(s,t) = \det \hat{R}(s,t)$ is obtained directly from the matrix $\hat{R}(s,t)$ given by (11) where the r_k are replaced by the $\hat{r}_k(n)$. Theorems 4.1 and 4.2 in Chapter 10 grant that *if X is gaussian the joint distributions of the $\hat{r}_0 \dots \hat{r}_k$, $\hat{\Phi}_1 \dots \hat{\Phi}_k$, $\hat{\rho}(s,t)$, $1 \leqslant s \leqslant k$, $0 \leqslant t \leqslant k$,* is asymptotically gaussian with the usual normalizing factor $\sqrt{n}$, and one can compute its asymptotic covariance matrix by Theorem 4.1 and 4.3 in Chapter 10. This implies the computations of partial derivatives like $(\partial/\partial r_j)[\rho(s,t)]$. The classical tool for this computation is the following elementary result: if a square matrix $R = (R_1 \dots R_s)$, where the R_s are column vectors, is a function of $y \in \mathbb{R}$, then one has

$$\frac{\partial}{\partial y}(\det R) = \det\left[\frac{\partial R_1}{\partial y}, R_2 \dots R_s\right] + \dots + \det\left[R_1 R_2 \dots \frac{\partial K_s}{\partial y}\right]$$

where each column is differentiated in turn.

The following particular case of asymptotic covariance computation is essential in practice.

Proposition (Quenouille). *Let X be a gaussian process with rational spectrum. Let $k,p \geqslant 0$. Consider the empirical estimator (based on n observations)*

$$\hat{v}(n) = [\hat{\Phi}_{p+1}(n), \dots, \hat{\Phi}_{p+k}(n)]$$

of the vector $v = (\Phi_{p+1}, \dots, \Phi_{p+k})$. Then if X is autoregressive with minimal order p, one has $v = 0$ and the joint distribution of $n\hat{v}(n)$ converges tightly as $n \to \infty$ toward the standard gaussian distribution $N(0,I)$ where I is the identity matrix of order k.

6.2. Covariances and Autocorrelations of High Order

For every stationary process with rational spectrum the covariances r_n tend to zero at exponential speed as $|n| \to +\infty$.

Indeed, the r_m are solutions of recurrence (2) for $m \geqslant q + 1$ and all the roots of $P(z)$ are of modulus > 1. The result is then a consequence of Section 1.1, Chapter 9.

We shall say that a process X with rational spectrum is **invertible** if its density $f(\lambda)$ has no zero on $\mathbb{T}$. This is equivalent to the assumption that *the canonical polynomial Q associated to X*, which a priori has its roots of moduli $\geqslant 1$, *has in fact no root of modulus* 1. *When X has rational spectrum and is invertible, the partial autocorrelations Φ_s tend to* 0 *at exponential speed as $s \to +\infty$.*

Indeed in this case the optimal predictor $\hat{X}_n$ of X_n at time $(n-1)$ may be written (cf. Chapter 9, Section 2.1)

$$\hat{X}_n = -\sum_{k \geqslant 0} d_{k+1} X_{n-k-1}$$

where

$$\frac{P}{Q}(z) = \sum_{k \geqslant 0} d_k z^k$$

and the d_k tend to zero at exponential speed as $k \to +\infty$. Let π be the projection of H^X on the space generated by $X_{n-1} X_{n-2} \dots X_{n-s}$. The linear regression $Y_{n,s}$ of X_n on $X_{n-1} \dots X_{n-s}$ is given by

$$Y_{n,s} = \pi(X_n) = \pi(\hat{X}_n) = -\sum_{k=1}^{s} d_k X_{n-k} - \pi\left[\sum_{k \geqslant s+1} d_k X_{n-k}\right]$$

whence

$$(14) \qquad \left\| Y_{n,s} + \sum_{k=1}^{s} d_k X_{n-k} \right\|_2 \leqslant \left(\sum_{k \geqslant s+1} |d_k| \right) \sqrt{r_0}\ .$$

But we have seen in 5.5 that

$$Y_{n,s} = -\sum_{k=1}^{s} \alpha_k(s) X_{n-k}$$

where α is given by (13). From (14) we conclude, setting $v_k = d_k - \alpha_k(s)$ and $v^* = (v_1 \dots v_s)$, that

$$(15) \qquad v^* \Gamma_s v \leqslant \chi^{2s}$$

where $0 < \chi < 1$ is a constant and $\Gamma_s = T_s(2\pi f)$ is the covariance matrix of $X_1 \dots X_s$. Since $f > 0$ is continuous, there is a $c > 0$ such that $f(\lambda) \geqslant c$ for all $\lambda \in \mathbb{T}$, and hence $\Gamma_s \geqslant T_s(2\pi c) = 2\pi c J$ where J_s is the identity matrix. But (15) then implies

$$\|v\|^2 \leqslant \frac{1}{2\pi c} \chi^{2s},$$

which a fortiori yields

$$|\Phi_s - d_s| \leqslant \frac{1}{\sqrt{2\pi c}} \chi^s$$

and concludes the proof.

6.3. Estimation of d for an ARIMA(p,d,q)

When a stationary process has rational spectrum and is invertible, the sequences Φ_k and $\rho_k = r_k/r_0$ must tend to 0 very fast as $k \to +\infty$. Start with a sequence of observations $Y_1...Y_N$ assumed to be already centered, and no a priori probabilistic information whatsoever. We want to fit a suitable ARIMA(p,d,q) model. Compute first the two estimated sequences $\hat{\Phi}_k$ and $\hat{\rho}_k$. *If we had* $d = 0$, these sequences should tend to 0 rather fast as $|k| \to \infty$. *If this is not the case*, we apply the same criterion to the derived series $\Delta Y_n = Y_n - Y_{n-1}$, then with $\Delta^2 Y_n$, and so on.

Thus we shall attempt to fit an ARMA(p,q) model to the series $\Delta^d Y$ *where* d *is a small integer such that at least one of the sequences* $\hat{\Phi}_k$, $\hat{\rho}_k$ associated to $\Delta^d Y$ tends to zero rather fast as $|k| \to \infty$. This demands the computation of confidence intervals for the $\hat{\rho}_k$ and $\hat{\Phi}_k$. In practice one avoids differentiations of order $d > 4$; if it seems necessary to differentiate further, one must question (a point which, in any case, is always pertinent) whether or not it would be useful to try an ARMA model with seasonal effects (cf. Section 9 below).

In order to decide that $d = \hat{d}$, at confidence level α (for instance $\alpha = 0.95$) one must be able to state that the ρ_ℓ associated to $\Delta^d Y$ are "zero" for $T \leqslant \ell \leqslant K$ with K as large as possible, **or that** that Φ_k are "zero" for $S \leqslant k \leqslant K$. Whence the necessity of simultaneous confidence intervals at level α for the ρ_ℓ, $T \leqslant \ell \leqslant K$ **or for** the Φ_k, $S \leqslant k \leqslant K$. See Section 7.2 for this computation.

7. Empirical Estimation of (p,q)

This stage is also called the **identification** of the model (see Section 10 and Chapter 14).

7.1. Approximate ARMA Models

Let X have rational spectrum and be of miniml type (p,q). We may write

$$X_n = \sum_{k \geqslant 0} c_k W_{n-k}$$

with $|c_k| \leqslant u^k$, where $u \in]0,1[$ is constant and W is the innovation of X. In particular this yields

$$\|X_n - \sum_{k=0}^{T} c_k W_{n-k}\| \leqslant \sigma^2 \sum_{k\geqslant T+1} |c_k|$$

and the MA(T) process

$$Y_n^{(T)} = \sum_{k=0}^{T} c_k W_{n-k}$$

verifies $\|X_n - Y_n^{(T)}\| \leqslant cu^T$ for all $n \in \mathbb{Z}$, where c is a constant. *Thus one can always approximate X as closely as wanted (in L^2) by* MA(T) *processes of order T large enough.* In fact as soon as $\sigma^2 c_{T_0}$ is negligible one may *practically consider X as an* MA(T_0). *Generally one has $q \leqslant T_0$ except for unusual configurations of coefficients* (this is only a heuristic remark).

If X is invertible, one has also

$$W_n = \sum_{k\geqslant 0} d_k X_{n-k}$$

with $|d_k| \leqslant u^k$ and $0 < u < 1$. The same argument as above shows that

$$\|W_n - \sum_{k=0}^{S} d_k X_{n-k}\|$$

tends to zero faster than cu^S as $S \to +\infty$. For S large, $\Sigma_{k=0}^{S} d_k X_{n-k}$ is practically a white noise with variance σ^2, and the process $U^{(S)}$ having rational spectrum and satisfying

$$\sum_{k=0}^{S} d_k U_{n-k}^{(S)} = W_n$$

is very close to X in L_2, uniformly in $n \in \mathbb{Z}$. Since $U^{(S)}$ is an AR(S) we see that *X can be approximated as closely as wanted (in L^2) by* AR(S) *processes of order S sufficiently large. Generally the value of S for which the approximation is acceptable are superior or equal to p (heuristic remark!) except for unusual configurations of coefficients.*

7.2. The Box-Jenkins Recipe to Obtain Upper Bounds for p and q

(a) We assume that, after a sufficient number of differentiations, we have replaced X by $Y = \Delta^d X$ such that at least one of the sequences $\hat{\rho}_k$, $\hat{\Phi}_k$ associated to Y tends to 0 when $k \to K$ with K "large".

(b) We want to test the hypothesis

$$H(S,T) = \{Y \text{ is approximately an AR}(S) \text{ and an MA}(T)\}$$

which by 7.1 is, for K large, approximately equivalent to

$$\tilde{H}(S,T) = \{\Phi_k \text{ and } \rho_\ell \text{ are approximately zero for}$$

$$K \geqslant k \geqslant S + 1 \text{ and } K \geqslant \ell \geqslant T + 1\}.$$

To test H at level $\alpha = 0.95$ or 0.99, we have to find ε, η such that

$$P_{H(S,T)}\Big\{\sup_{S+1 \leqslant k \leqslant K} |\hat{\Phi}_k - \Phi_k| \leqslant \varepsilon \text{ and } \sup_{T+1 \leqslant \ell \leqslant K} |\hat{\rho}_\ell - \rho_\ell| \leqslant \eta\Big\} \geqslant \alpha.$$

We shall then accept $H(S,T)$ if

$$\sup_{S+1 \leqslant k \leqslant K} |\hat{\Phi}_k| \leqslant \varepsilon \quad \text{and} \quad \sup_{T+1 \leqslant \ell \leqslant K} |\hat{\rho}_\ell| \leqslant \eta.$$

It is obviously sufficient to grant

$$(16) \quad \begin{aligned} &P_{H(S,T)}\Big[\sup_{S+1 \leqslant k \leqslant K} |\hat{\Phi}_k - \Phi_k| > \varepsilon\Big] \leqslant \frac{1}{2}(1 - \alpha) \\ &P_{H(S,T)}\Big[\sup_{T+1 \leqslant \ell \leqslant K} |\hat{\rho}_\ell - \rho_\ell| > \varepsilon\Big] \leqslant \frac{1}{2}(1 - \alpha). \end{aligned}$$

The last inequality is true as soon as

$$(17) \quad P[|\hat{\rho}_k - \rho_k| > \eta] \leqslant \frac{1 - \alpha}{2(K - T)} \qquad T + 1 \leqslant k \leqslant K.$$

Since the $\hat{\Phi}_k$ are asymptotically independent (under hypothesis $H(S,T)$) for $k \geqslant S + 1$, then (16) is true as soon as

$$(18)\qquad P_{H(S,T)}[|\hat{\Phi}_k - \Phi_k| \leq \varepsilon] \geq \left[1 - \frac{1}{2}(1-\alpha)\right]^{1/K-S} \quad \text{for } S+1 \leq k \leq K.$$

In the computations of ε and η determined by (16) and (18), one uses the asymptotic results stated above, concerning $\hat{\Phi}_k$, $\hat{\rho}_k$. The asymptotic variances of the $\hat{\rho}_k$ involve the ρ_j, $j \in \mathbb{N}$ which are unknown. As a first approximation, in this "error computation" one replaces brutally the ρ_j by the $\hat{\rho}_j$, to estimate the asymptotic variances of the $\hat{\rho}_k$. Note that under $H(S,T)$ one has $\rho_j = 0$ for $j > T$.

Once $H(S,T)$ is accepted, with S, T as small as possible, the arguments of 7.1 give weight to the conclusion that the minimal type (p,q) of Y "should" satisfy $p \leq S$, $q \leq T$. At this stage, Box and Jenkins propose generally to consider separately several or even all the (p,q) pairs bounded by (S,T). It seems (theoretically at least!) more rigorous to apply the "corner method", still taking account of the information $p \leq S$, $q \leq T$.

7.3. The Corner Method ([7])

One computes the estimates $\hat{\rho}(s,t)$ of the determinants $\rho(s,t)$ introduced in (12). We are looking for p,q such that the $\hat{\rho}(s,t)$ are zero for $\{s \geq p+1,\ t \geq q+1\} \cup \{s \geq p+1,\ t = q\}$. In principle, it would be sensible to construct *simultaneous* confidence intervals for the $\hat{\rho}(s,t)$, but one can often obtain useful information by direct inspection of the array $\hat{\rho}(s,t)$. In practice, it is better to normalize the $\hat{\rho}(s,t)$ by $1/\hat{r}_0^s$, that is to compute the $\hat{\rho}(s,t)$ starting from the correlations $\hat{\rho}_k$ instead of the covariances $\hat{r}_k$, since one has $|\rho(s,t)/r_0^s| \leq 1$.

7.4. Conclusion

Except in favorable cases, the methods of 7.2 and 7.3 often fail to provide a clear cut estimation of p, d, q, in particular for low values of n. This empirical study generally ends up with a finite number of possibilities for the triple p, d, q, which will have to be considered separately (efficient estimation of coefficients, cf. Chapter 12). The choice between the possibilities raises several statistical problems which will be considered in Chapter 15 (quality of fit, etc. ...).

8. Complement: A Direct Method of Computation for the b_k

Let X be a process with rational spectrum, having minimal type (p,q). Assume that the covariances of X are given. We can then compute directly the a_k, $1 \leq k \leq p$, by solving the Yule-Walker equations.

Consider then the process Y given by

$$Y_n = \sum_{k=0}^{p} a_k X_{n-k} .$$

The covariances s_j of Y are explicitly given by (6). The process Y has $(0,q)$ minimal type and satisfies the canonical MA(q) relation

$$Y_n = \sum_{\ell=0}^{q} b_\ell W_{n-\ell}$$

where W is the innovation of X, and the innovation of Y as well. The spectral density f of Y may be written, with $z = e^{-i\lambda}$

$$f(\lambda) = \frac{\sigma^2}{2\pi}|Q(z)|^2 \qquad \lambda \in \mathbb{T} .$$

We are going to approximate Y by an AR(S) with S large (cf. Section 7.1). Thus we want to find a polynomial P_S of degree S such that

$$f(\lambda) \simeq \frac{\sigma^2}{2\pi}\left|\frac{1}{P_S(z)}\right|^2 \qquad \text{with } z = e^{-i\lambda}, \ \lambda \in \mathbb{T}.$$

The computation of the best approximation P_S is obvious: we simply have to solve *the Yule-Walker equations of order* S (those which are used to compute the partial autocorrelations) *written for the process* Y, *whose covariances* s_j *are known and are in fact equal to zero when* $|j| \geq q$. *This can be done explicitly and we see that* $1/P_S(z) \times Q(z)$ *tends to* 1 *as* $S \to +\infty$. We thus obtain a good approximation $Q_S(z)$ of $Q(z)$ by computing the expansion of $1/P_S(z)$ as a formal series, which may be done explicitly thanks to the following formula (which converges for S large)

$$\frac{1}{P_S} = \frac{1}{1+P_S-1} = 1 - (P_S-1) + (P_S-1)^2 + \dots + (-1)^k(P_S-1)^k + \dots$$

This approximation $Q_S(z)$ is a *series* in z^k, $k \geq 0$, whose coefficients $u_k(S)$ converge towards those of Q for $0 \leq k \leq q$ and towards 0 for $k \geq q + 1$.

With respect to the classical methods presented in Section 3 (in improved versions!) the method just outlined *presents only advantages for* $q \geq 4$.

9. The ARMA Models with Seasonal Effects

When it seems difficult to modelize by an ARIMA(p,d,q) *with d small*, one may try to fit an ARMA model with seasonal effects. By graph inspection, available a priori information, vague botanical information on correlation-autocorrelation patterns or typical prediction curves supplied by an ad hoc catalogue of seasonal ARMA models (see [10] for instance), *we select a priori a* **small** *number of periods* $s_1 s_2 \dots s_k$ ($k \leq 3$ is strongly suggested, and in most cases it is best to limit oneself to $k = 2$, with $s_1 = 1$ and s_2 adequately selected). One computes the estimated correlations and partial autocorrelations for the processes $A_R Y$ where

$$R(z) = (1 - z^{s_1})^{d_1} \dots (1 - z^{s_k})^{d_k},$$

letting the multi-index $(d_1 \dots d_k)$ increase methodically, starting with $(0 \dots 0)$ and keeping the d_j small. Here too the bound $d_j \leq 3$ is strongly suggested.

One stops as soon as one is satisfied that $A_R Y$ is reasonably likely to have rational spectrum, that is if at least one of the two sequences $\hat{\rho}_k$, $\hat{\Phi}_k$ tends to 0 reasonably fast when $k \to +\infty$. The situation is quite similar to the estimation of d in Section 6, but now $d = (d_1 \dots d_k)$ is a *multi-index*, which does not change anything to the arguments used in Section 6.

Once the multi-index d has been selected, one goes on exactly as above, to select (p,q) starting with $A_R Y$.

10. *A Technical Result: Characterization of Minimal Recursive Identities*

We are going to prove the classical lemma 5.3 since the references [7] and [22] quoted above only cover 5.3(i),(ii) while we have needed 5.3(iii),(iv).

(a) *If* (r_n) *is a numerical sequence verifying*

$$\alpha_0 r_t + \dots + \alpha_p r_{t-p} = 0 \qquad t \geq q + 1$$

then for $s \geq p + 1$, $t \geq q + 1$ the columns $R_j(s,t)$ of $R(s,t)$ are linked by $\alpha_0 R_1 + \dots + \alpha_p R_{p+1} = 0$, and hence $\rho(s,t) = 0$ for $s \geq p + 1$, $t \geq q + 1$.

(b) If (r_n) is a sequence such that $\rho(m + 1,t) = 0$, $\rho(m,t) \neq 0$ with m,t fixed, there exists a nonzero vector β such that

$$(19) \qquad \beta_0 R_1(m + 1,t) + \dots + \beta_m R_{m+1}(m + 1,t) = 0.$$

Since $\rho(m,t) \neq 0$, β_0 cannot be zero, because the columns of $R(m,t)$ would then be linked by $\beta_1 R_1(m,t) + \dots + \beta_m R_m(m,t) = 0$ and we would have $\rho(m,t) = 0$. Hence we may take $\beta_0 = 1$, and by combination of columns, (19) implies $\rho(m,t) = \pm\beta_m^m \rho(m,t+1)$ whence $\rho(m,t+1) \neq 0$. An analogous computation yields $\rho(m,t-1) \neq 0$.

Thus for every sequence (r_n), for all *fixed* pairs m, t of integers, the relations $\rho(m+1,t) = 0$ and $\rho(m,t) \neq 0$ imply $\rho(m,t-1) \neq 0$ and $\rho(m,t+1) \neq 0$.

(b*) By an analogous argument on *lines* one shows that for s,m *fixed* and arbitrary the relations $\rho(s,m+1) = 0$ and $\rho(s,m) \neq 0$ imply $\rho(s-1,m) \neq 0$ and $\rho(s+1,m) \neq 0$.

(c) Assume that (r_n) satisfies $\rho(p+1,t) = 0$ for $t \geq q + 1$ and $\rho(p+1,q) \neq 0$. By (b) one has $\rho(p,t) = 0$ for all $t \geq q$. Starting from $\rho(p+1, q+1) = 0$ and $\rho(p,q+1) \neq 0$ we obtain as in (b) coefficients $\alpha_0 \dots \alpha_p$ with $\alpha_0 = 1$ such that the relation

$$(20) \qquad \alpha_0 R_1(p + 1,t) + \dots + \alpha_p R_{p+1}(p + 1,t) = 0$$

be true for $t = q + 1$. Assume now that (20) holds for $q + 1 \leq t \leq u$, which is equivalent to

$$(21) \qquad \varphi(n) = r_n + \alpha_1 r_{n-1} + \dots + \alpha_p r_{n-p} = 0$$

for $q + 1 \leq n \leq u + p$.

By linear combination of columns, (20) written for $t = u$ yields directly

$$\rho(p + 1,u + 1) = (-1)^{p+1}\varphi(u + p + 1)\rho(p,u)$$

whence $\varphi(u + p + 1) = 0$. By induction on n one has thus shown that $\varphi(n) \equiv 0$ for $n \geqslant q + 1$ and r_n satisfies a (p,q)-recursive relation.

(c*) Same type of argument on lines, with the use of (b*) to see that if $\rho(s,q + 1) = 0$ for $s \geqslant p + 1$, and $\rho(p,q + 1) \neq 0$ then (r_n) verifies a (p,q)-recursion.

(d) Start with a sequence (r_n) satisfying a minimal (p,q)-recursive relation. By (a) we have $\rho(s,t) = 0$ for $s \geqslant p + 1$, $t \geqslant q + 1$. If one of the $\rho(p,t)$, $t \geqslant q + 1$ were equal to zero, then all of them would be zero by (b), and by (c), the r_n would satisfy a $(p - 1,q)$ recursive relation, which is impossible. Hence $\rho(p,t) \neq 0$ for $t \geqslant q + 1$. Using (b*) and (c*) one concludes similarly that $\rho(s,q) \neq 0$ for $s \geqslant p + 1$. *Thus in statement* 5.3 *we have proved that* (i) *implies* (ii).

Conversely when (ii) is true, the sequence (r_n) satisfies a (p,q)-recursive relation by (a). Assume that it satisfies another (p',q')-recursive relation. If we had $p' < p$, then $\rho(p,t)$ would be zero for t large by (a), contradicting (ii). If we had $q' < q$, then $\rho(p' + 1,q) = 0$ by (a), contradicting (ii). Whence $p' \geqslant p$, $q' \geqslant q$, and (ii) $\Rightarrow$ (i), *which yields in* 5.3 *the equivalence* (i) $\Leftrightarrow$ (ii).

Assume (iii) to be true. By (c) the sequence (r_n) satisfies a (p,q)-recursive relation. By (a) one has then $\rho(s,t) = 0$ for $s \geqslant p + 1$, $t \geqslant q + 1$. Thanks to (b), one proves by induction on s, then by induction on t that $\rho(p,t) \neq 0$ for $t \geqslant q + 1$ and $\rho(s,q) \neq 0$ for $s \geqslant p + 1$. Whence (iii) $\Rightarrow$ (ii) and hence, the converse implication being trivially true, (iii) $\Leftrightarrow$ (ii). Using (a), (b*) and (c*) instead of (a), (b) and (c) one sees similarly that (ii) $\Leftrightarrow$ (iv) in 5.3.

11. Empirical Estimation and Identification

A few words about our use of the term "empirical", which is more specific than in standard literature on the subject.

"Empirical" is here associated to a notion of estimation valid under weak hypotheses on the model. In the case of second order stationary processes, one may consider three *levels* of models (sets of process types).

(a) **Nonparametric models (wide sense)**: such as the set of all regular, second-order stationary processes, having a spectral density.

(b) **Parametrized models (wide sense)**: for instance the set of all ARMA(p,q) processes with p, q fixed. The processes are parametrized here by the coefficients a, b of the ARMA equation and the variance of the innovation (which, since $a_0 = b_0 = 1$, gives a $(p + q + 1)$-dimensional parameter).

c) **Parametrized models (strict sense)**: same as in b) but with a supplementary hypothesis such as "the process is gaussian." The parametrization is unchanged, but the joint distributions are *specified*.

If $(X_1 \dots X_n)$ is the observation of a second-order stationary process, *the empirical distribution*

$$\hat{F}_n = \frac{1}{n} \sum_{i=1}^{n} \delta_{X_i}$$

is an estimate of the distribution F of X_1. Here δ_a is the Dirac mass at point a.

More generally if ℓ, $h_1 < h_2 < \dots < h_\ell$ are positive integers, *the empirical distribution*

$$\hat{F}_{n,h_1 \dots h_\ell} = \frac{1}{n - h_\ell} \sum_{i=1}^{n-h_\ell} \delta_{Z_i}$$

where

$$Z_i = (X_i, X_{i+h_1}, \dots, X_{i+h_\ell}) \in \mathbb{R}^{\ell+1}$$

is an estimate of the law $F_{h_1 \dots h_\ell}$ of Z_1, by the finite (random) subset $(Z_1 \dots Z_{n-h_\ell})$ of $\mathbb{R}^{\ell+1}$.

The "parameters" which can be estimated empirically are those which are defined by sufficiently smooth functionals of the joint distributions $F_{h_1 \dots h_\ell}$. Thus the mean $\int x \, dF(x)$ is naturally estimated by

$$\int_{\mathbb{R}} x \, d\hat{F}_n(x) = \frac{1}{n} \sum_{i=1}^{n} X_i,$$

and the covariance

$$r_k = \int_{\mathbb{R}^2} x_1 x_{1+k} dF_k(x_1, x_{1+k})$$

by the empirical covariance

$$\hat{r}_k = \int_{\mathbb{R}^2} x_1 x_{1+k} d\hat{F}_{n,k}(x_1, x_{1+k}) = \frac{1}{n-k} \sum_{i=1}^{n-k} X_i X_{i+k} \,.$$

Similarly for an ARMA(p,q) process, the coefficients $a = \psi(r_0 ... r_k)$ and $b = \eta(r_0 ... r_k)$ of the ARMA canonical form are estimated empirically by $\hat{a} = \psi(\hat{r}_0 ... \hat{r}_k)$, $\hat{b} = \eta(\hat{r}_0 ... r_k)$.

The order $(\hat{p},\hat{q})$ selected, as sketched above, by systemtic study of the empirical correlations and partial autocorrelations is also an empirical estimator of (p,q).

The following chapters consider models parametrized in the strict sense (guassian processes) for which more accurate estimators than the empirical estimators can be found.

Bibliographical Hints

The distinction between empirical and efficient estimators is often blurred in the literature.

Numerous papers have been concerned with the asymptotic study of empirical estimators (in particular, *T. W. Anderson* and *Quenouille*). *Box* and *Jenkin's* book has widely imposed the use of empirical correlations and partial autocorrelations to determine the AR and MA orders (see also *Begin, Gourieroux, Montfort* for the corner method). *Box-Jenkins* include numerous interesting heuristic considerations, and decidely avoid the use of simultaneous confidence regions for multidimensional parameters. They prefer to use quick qualitative graphical comparisons between empirical and theoretical quantities.

Chapter XII
EFFICIENT ESTIMATION FOR THE PARAMETERS OF A PROCESS WITH RATIONAL SPECTRUM

1. Maximum Likelihood

1.1. Definitions. Let X be a real valued random process with discrete time, whose distributions P_θ depends on the parameter $\theta \in \Theta$, where Θ is a subset of $\mathbb{R}^k$. Assume that the joint distribution of the observations $X_1 ... X_N$ under P_θ has a density $F_{\theta,N}(x_1 ... x_N)$ with respect to Lebesgue measure on $\mathbb{R}^N$.

Given $X_1(\omega) ... X_N(\omega)$ one can try to estimate θ by points θ of Θ which maximize the probability $P_\theta[X_1 ... X_N \in V]$ where V is a very small neighborhood of $X_1(\omega) ... X_N(\omega)$. This suggests to seek $\hat{\theta}_N(X_1 ... X_N)$ such that

$$\sup_{\theta \in \Theta} F_{\theta,N}(X_1 \dots X_N) = F_{\hat{\theta}_N,N}(X_1 \dots X_N) .$$

When such a $\hat{\theta}_N$ exists, one says that $\hat{\theta}_N$ is **a maximum likelihood estimator for θ**. In practice, one often maximizes approximations of $F_{\theta,N}$, which yields approximate maximum likelihood estimators. *Under restrictive, but reasonable hypotheses* (cf. Chapter 13) these estimators have an excellent asymptotic behaviour: the distribution of $\sqrt{n}(\hat{\theta}_N - \theta)$ is asymptotically gaussian, centered, with *minimal* asymptotic covariance matrix. Such estimators $\hat{\theta}_N$ are said to be **efficient**.

1.2. The Likelihood of Stationary Gaussian Processes

Let X be a centered, regular, stationary gaussian process, not identically zero. Let f be its spectral density. The random vector

$$X(n) = \begin{bmatrix} X_1 \\ \cdot \\ \cdot \\ X_n \end{bmatrix}$$

has covariance matrix $\Gamma_n = T_n(2\pi f)$ where T_n is the Toeplitz matrix of order n and Γ_n is invertible (cf. Chapter 10, Section 2.3). The gaussian vector $X(n)$ thus has on $\mathbb{R}^n$ the density

$$f_n(x) = (2\pi)^{-n/2}(\det \Gamma_n)^{-1/2}\exp\left[-\frac{1}{2}x^*\Gamma_n^{-1}x\right], \quad x \in \mathbb{R}^n.$$

Call **log likelihood** of $X_1 \ldots X_n$ the random variable

$$L_n(f, X_1 \ldots X_n) = \log f_n(X_1 \ldots X_n)$$

given by

$$L_n(f, X_1 \ldots X_n) = -\frac{1}{2}[n \log 2\pi + \log \det T_n(2\pi f)$$
$$+ X(n)^*[T_n(2\pi f)]^{-1}X(n)].$$

Let W be the innovation of X, and σ^2 the variance of W. One knows (Chapter 4, Section 3) that

$$X_n = \sum_{k \geqslant 0} c_k W_{n-k} \quad \text{where} \quad \sum_{k \geqslant 0} | c_k|^2 < \infty.$$

This implies

(2) $$X(n) = \sum_{-\infty < k \leqslant n} M_k W_k$$

where the M_k are deterministic $(n,1)$ matrices, computed in an obvious way from the c_j, $j \geqslant 0$. Let $\hat{W}_k = E[W_k | X(n)]$ be the linear regression of W_k on $X_1 \ldots X_n$. One may write

$$\hat{W}_k = G_k^* X(n) = X(n)^* G_k$$

where G_k is a deterministic $(n,1)$ matrix. From (2) one

deduces that

$$(3)\quad \begin{aligned} \sigma^2 M_k &= E[X(n)W_k] = E[X(n)\hat{W}_k] = E[X(n)X(n)^*G_k] \\ &= \Gamma_n G_k, \quad k \leqslant n. \end{aligned}$$

On the other hand (2) implies

$$\Gamma_n = E[X(n)X(n)^*] = E\left[\left(\sum_{-\infty<k\leqslant n} M_k W_k\right)\left(\sum_{-\infty<\ell\leqslant n} W_\ell M_\ell^*\right)\right]$$

$$= \sigma^2 \sum_{-\infty<k\leqslant n} M_k M_k^*$$

whence by (3)

$$\sigma^2 \Gamma_n = \sum_{-\infty<k\leqslant n} \Gamma_n G_k G_k^* \Gamma_n$$

and after left- and right-multiplication by Γ_n^{-1}

$$\sigma^2 \Gamma_n^{-1} = \sum_{-\infty<k\leqslant n} G_k G_k^*$$

which finally yields

$$(4)\quad \begin{aligned} X(n)^*\Gamma_n^{-1}X(n) &= \frac{1}{\sigma^2} \sum_{-\infty<k\leqslant n} X(n)^* G_k G_k^* X(n) \\ &= \frac{1}{\sigma^2} \sum_{-\infty<k\leqslant n} |\hat{W}_k|^2. \end{aligned}$$

Moreover we have seen (Chapter 4, Section 4) that the one step error of prediction σ^2 satisfies

$$\log \sigma^2 = \lim_{n\to\infty} \frac{1}{n} \log \det T_n(2\pi f).$$

This suggests the replacement of L_n by the **approximation** $\tilde{L}_n$ given by

$$(5)\qquad \tilde{L}_n(f, X_1 \dots X_n) = -\frac{1}{2}[n \log 2\pi\sigma^2 + X(n)^*[T_n(2\pi f)]^{-1}X(n)]$$

which, due to (4), may also be written as

$$(5')\qquad \tilde{L}_n(f, X_1 \dots X_n) = -\frac{1}{2}\left[n \log 2\pi\sigma^2 + \frac{1}{\sigma^2} \sum_{-\infty<k\leqslant n} |\hat{W}_k|^2\right].$$

For $\varphi \in L^1(\mathbb{T})$, having Fourier coefficients $\hat{\varphi}_k$ we have defined $\alpha(\varphi)$ (cf. Chapter 10, Section 3) by

$$\alpha(\varphi) = \sum_k (1 + |k|)|\hat{\varphi}_k|.$$

Assume that f satisfies $\alpha(\log f) < \infty$, which is for instance the case if $0 < m \leqslant f \leqslant M$ and $|f'''| \leqslant M$, where m, M are constants. We shall see in Chapter 13 that we then have

$$\text{(6)} \qquad \sup_{n \geqslant 1} |\log \det T_n(2\pi f) - n \log \sigma^2| \leqslant c(f)$$

where the constant $c(f)$ only depends on $\alpha(\log f)$. This implies

$$\text{(7)} \qquad \sup_{n \geqslant 1} |L_n(f, X_1 \ldots X_n) - \tilde{L}_n(f, X_1 \ldots X_n)| \leqslant \frac{1}{2} c(f)$$

by (1) and (5). Since $X(n)$ is gaussian and has covariance $\Gamma_n = T_n(2\pi f)$ the vector $Y_n = \Gamma_n^{-1/2} X(n)$ is standard gaussian, and $X(n)^* \Gamma(n)^{-1} X(n) = Y_n Y_n^*$ is the sum of the squares of n independent standard gaussian variables. Hence $X(n)^*[T_n(2\pi f)]^{-1} X(n)$ has the law of a χ^2 with n degrees of freedom, and in particular has mean n and variance $2n$. This suggests that L_n is of size $n \times$ (constant), and the passage from L_n to $\tilde{L}_n$ should have no asymptotic impact. This point of view will be rigorously justified in Chapter 13.

1.3. Parametrization of Gaussian Processes with Rational Spectrum

Let X be centered, invertible, stationary gaussian process with rational spectrum. Let us denote by

$$X_n + a_1 X_{n-1} + \ldots + a_p W_{n-p} = W_n + b_1 W_{n-1} + \ldots + b_q W_{n-q}$$

the minimal *canonical* ARMA relation linking X and its innovation W, and let σ^2 be the variance of W. *Assume p and q to be known.* We want to estimate $\theta = (a_1 \ldots a_p, b_1 \ldots b_q, \sigma) \in \mathbb{R}^{p+q+1}$. The polynomials $P(z) = 1 + \ldots + a_p z^p$ and $Q(z) = 1 + \ldots + b_q z^q$ must have all their roots of modulus > 1, with $a_p b_q \neq 0$ and Q/P must be irreducible. This constrains (a,b) to belong to an open subset $\mathcal{U}$ of $\mathbb{R}^{p+q}$. In principle, then θ belongs to the open set $\mathcal{U} \times \mathbb{R}^+ \subset \mathbb{R}^{p+q+1}$.

But in practice it is *essential* to restrict a priori P and Q to have all their roots of modulus $\geqslant 1 + \varepsilon$, with for instance $\varepsilon = .01$. On the other hand the simplest asymptotic results deal with the case of a compact parameter space. Hence we will assume here that *the parameter space* Θ *is a compact subset of* $\mathcal{U} \times (\mathbb{R}^+ - \{0\})$. Such parameter spaces may be called *canonical compact parameter spaces* and they are obviously such that the map $\theta \to f_\theta$ is injective.

A simple example of such a parameter space is obtained by imposing on (P,Q,σ) the following restrictions: $\alpha \leqslant \sigma \leqslant \beta$, $a_0 = b_0 = 1$, $|a_p| \geqslant \varepsilon$, $|b_q| \geqslant \varepsilon$, and for any roots a of P and η of Q one assumes $|\tau| \geqslant 1 + \varepsilon$, $|\eta| \geqslant 1 + \varepsilon$, $|\tau - \eta| \geqslant \varepsilon$ where $\varepsilon > 0$, $\alpha > 0$, $\beta > \alpha$, are fixed and arbitrary. More general parameterizations will be considered in Section 3 below and in Chapter 13.

1.4. Theorem. *Let* $\Theta \subset \mathbb{R}^{p+q+1}$ *be as in* 1.3. *Let* $X_1 ... X_n$ *be* n *observations of an invertible, stationary, centered gaussian process* X, *with rational spectrum parametrized by* $\theta \in \Theta$. *Let* f_θ *be the spectral density of* X *and let* $\widetilde{L}_n(\theta, X_1 ... X_n) = \widetilde{L}_n(f_\theta, X_1 ... X_n)$ *be the approximate log-likelihood of* X *given by* (5) *and* (5'). *Let* $\widetilde{\theta}_n$ *be any approximate maximum likelihood estimator of* θ *such that*

$$\widetilde{L}_n(\widetilde{\theta}_n) = \sup_{\theta \in \Theta} \widetilde{L}_n(\theta).$$

Then, whenever θ *is an* **interior** *point of* Θ, *the estimator* $\widetilde{\theta}_n$ *is a consistent and efficient estimator of* θ, *and the law of* $\sqrt{n}(\widetilde{\theta}_n - \theta)$ *converges tightly as* $n \to \infty$ *to the gaussian law* $N(0, J(\theta)^{-1})$ *where* $J(\theta)$ *is the* **asymptotic information matrix** (cf. *Chapter* 13) *given by*

$$(8) \qquad J_{k\ell}(\theta) = \frac{1}{4\pi} \int_{-\pi}^{\pi} \frac{1}{f_\theta^2(\lambda)} \frac{\partial f_\theta}{\partial \theta_k}(\lambda) \frac{\partial f_\theta}{\partial \theta_\ell}(\lambda) d\lambda.$$

Proof. It will be given in Chapter 13, Theorem 4.4. We shall see in Section 3 that J_θ *is indeed invertible* and we shall give an *explicit expresssion* of $J(\theta)$.

1.5. A Strategy to Compute $\hat{\theta}_n$

For the practical computation of $\hat{\theta}_n$, with $X_1 \dots X_n$ given, we shall use (5') which will be rewritten as

$$(9) \qquad -2\hat{L}_n(\theta) = n \log(2\pi\sigma^2) + \frac{1}{\sigma^2} F_n$$

where

$$F_n = \sum_{-\infty<k\leq n} [E(W_k|X_1 \dots X_n)]^2 = \sum_{-\infty<k\leq n} |\hat{W}_k|^2.$$

A priori F_n depends on $X(n)$ and $\theta = (a,b,\sigma)$. But since $E[W_k|X(n)]$ is a linear function $h[X(n)]$ we have for all $u \in \mathbb{R}$,

$$E[uW_k|uX(n)] = E[uW_k|X(n)] = uh(X(n)) = h[uX(n)]$$

which proves that F_n *is independent of* σ and we shall write $F_n = F_n(a,b,X(n))$.

To maximize $\hat{L}_n(\theta)$ in θ, we can then take the following two steps. First, one seeks $\tilde{a}$, $\tilde{b}$ such that

$$F_n(\hat{a},\hat{b},X(n)) = \inf_{(a,b)\in U} F_n(a,b,X(n))$$

where $\Theta = U \times \mathbb{R}^+$ as above. One minimizes then, in $\sigma > 0$, the expresssion

$$n \log 2\pi\sigma^2 + \frac{1}{\sigma^2} F_n[\hat{a},\hat{b},X(n)]$$

which immediately gives (computing the zero of the derivative in σ)

$$(10) \qquad \hat{\sigma}^2 = \frac{1}{n} F_n[\hat{a},\hat{b},X(n)].$$

The delicate phase is hence the computation of $\hat{a}$, $\hat{b}$ and the sought for estimator is $\tilde{\theta}_n = (\tilde{a},\tilde{b},\tilde{\sigma})$.

2. The Box-Jenkins Method to Compute $(\hat{a},\hat{b})$

Since no explicit simple form is available for $F_n(a,b,X(n))$, the minimization in (a,b) is done by exploration of the graph of $(a,b) \to F_n(a,b,X(n))$. One begins with the computation, by the methods outlined in Chapter 11, of the empirical estimators

$(\hat{a},\hat{b})$ of (a,b). One selects then a finite network $\mathcal{R}$ of points of $\mathbb{R}^{p+q}$ in the neighborhood of $(\hat{a},\hat{b})$. By the "backforecasting method" described below, one computes the value $F_n(\cdot,\cdot,X(n))$ at each point of $\mathcal{R}$. The minimizing argument $(\tilde{a},\tilde{b})$ is then obtained by direct inspection.

In practice for $p + q \geqslant 5$ it is faster to seek the minimum of F_n on $\mathcal{R}$ by one of the many gradient methods (cf. [53]), which amount roughly to move the argument in successive steps, each time in the direction opposed to the gradient of F_n.

This requires the computation of the $\partial F_n/\partial a_k$, $\partial F_n/\partial b_\ell$ by naive approximations like

$$\frac{\partial F_n}{\partial a_k} \simeq \frac{1}{\varepsilon}[F_n[a + \varepsilon e_k,b,X(n)] - F_n[a,b,X(n)]$$

with ε small and e_k adequate basis vector in $\mathbb{R}^{p+q}$.

It is wise to visualize as clearly as possible the behaviour of F_n around its minimum, and to check graphically the local convexity of F_n around $(\tilde{a},\tilde{b})$, which is "equivalent" to checking that the minimum is clearly defined.

Indeed when the minimum of F_n is practically reached for a whole family of points in $\mathbb{R}^{p+q}$ which is not concentrated in a very small neighborhood of $(\tilde{a},\tilde{b})$, *one can guess that the dimension* $(p + q + 1)$ *of the model has been overestimated* and it is strongly suggested to start again the estimation process with smaller p, q.

2.1. Computation of $F_n(a,b,X(n))$ for a, b Given

The **backforecasting method** which we are about to describe is due to Box-Jenkins. It is efficient when the roots of the polynomials P, Q associated to (a,b) are of *modulus clearly larger than* 1.

Adopt the notation $\hat{Y} = E[Y|X(n)]$ for all $Y \in L^2(\Omega,P)$. By Chapter 9, Section 3, there exists a white noise U having variance σ^2, and having the *same future* as X, which satisfies

$$\sum_{k=0}^{p} a_k X_{m+k} = \sum_{\ell=0}^{q} b_\ell U_{m+\ell}, \qquad m \in \mathbb{Z}$$

$$\sum_{k=0}^{p} a_k X_{m-k} = \sum_{\ell=0}^{q} b_\ell W_{m-\ell}, \qquad m \in \mathbb{Z}.$$

Taking conditional expectations, we obtain

$$(11) \qquad \sum_{k=0}^{p} a_k \hat{X}_{m+k} = \sum_{\ell=0}^{q} b_\ell \hat{U}_{m+\ell}, \qquad m \in \mathbb{Z}$$

$$(12) \qquad \sum_{k=0}^{p} a_k \hat{X}_{m-k} = \sum_{\ell=0}^{q} b_\ell \hat{W}_{m-\ell}, \qquad m \in \mathbb{Z} .$$

(a) Start with an arbitrary vector $\alpha = (\alpha_1 ... \alpha_q) \in \mathbb{R}^q$ and begin the iterative estimation of $\hat{X}$, $\hat{U}$, $\hat{W}$ by arbitrarily letting

$$(13) \qquad \hat{U}_{n-p+j} = \alpha_j, \qquad 1 \leqslant j \leqslant q.$$

The relation (11) used with m decreasing from $n - p$ to 1 allows then the successive computation of the $\hat{U}_m$ for $n - p \geqslant m \geqslant 1$; indeed the $\hat{X}_j$ involved in this computation have indices $j \in [1,n]$ and one clearly has,

$$(14) \qquad \hat{X}_j = E(X_j | X_1 ... X_n) = X_j \qquad \text{for } 1 \leqslant j \leqslant n.$$

On the other hand, the U_k, $k \leqslant 0$ are orthogonal to the U_m, $m \geqslant 1$, and hence to the X_m, $m \geqslant 1$ since X and U have the same future, so that

$$(15) \qquad \hat{U}_k = 0 \qquad \text{for } k \leqslant 0.$$

At the end of phase (a), starting with the initial point α, and with a, b, $X(n)$, all the $\hat{U}_m$, $m \leqslant n - p + q$, have been computed.

(b) Relation (11) used for $m = 0,-1,-2, ...$ then successively gives all the $\hat{X}_m$, $m \leqslant 0$. For $m \leqslant 0$ and $|m|$ large enough, (11) and (15) imply

$$\sum_{k=0}^{p} a_k \hat{X}_{m+k} = 0.$$

Since $P(z)$ has all its roots of modulus > 1, the $\hat{X}_m$ must converge to zero (cf. Chapter 9, Section 11) at exponential speed as $m \to -\infty$. Beyond an index $(-S)$ which depends on the accuracy of actual computations, the $\hat{X}_m$, $m \leqslant -S$ are practically zero.

(c) Since

$$W_m = \sum_{k \geqslant 0} d_k X_{m-k}$$

with $\Sigma_{k\geqslant 0}|d_k| < \infty$, the

$$\hat{W}_m = \sum_{k\geqslant 0} d_k \hat{X}_{m-k}$$

will also be practically zero for $m \leqslant -S$. One sets hence $\hat{W}_m = 0$ for $m \leqslant -S$. The $\hat{X}_k$ being now known for $k \leqslant n$, one uses (12), with m increasing from $(-S+1)$ to n, to successively compute the $\hat{W}_m$, $-S+1 \leqslant m \leqslant n$. On the other hand, W and X having the same past, the W_k, $k \geqslant n+1$, are orthogonal to the X_m, $m \leqslant n$ and hence

$$\hat{W}_k = 0 \qquad \text{for } k \geqslant n+1.$$

At the end of stage (c) one has obtained all the $\hat{W}_m$, $m \in \mathbb{Z}$ in terms of α, a, b, and $X(n)$.

(d) One uses (12) with $m = n+1, n+2, \ldots$ to successively compute the $\hat{X}_m$, $n \geqslant m+1$. As in (b) one proves that the $\hat{X}_m$ tend to zero at exponential speed when $m \to +\infty$. Thus there is an integer T such that the $\hat{X}_m$ are practically zero for $m \geqslant T$. At the end of stage (d) one has obtained all the $\hat{X}_m$, $m \in \mathbb{Z}$.

(e) As in (c) one sees that the $\hat{U}_k$ are practically zero for $k \geqslant T$. One *sets* hence $\hat{U}_k = 0$ for $k \geqslant T$, and uses (11) with m decreasing from T, to successively compute the $\hat{U}_m$, $T \geqslant m \geqslant n-p+1$. *This gives us a new set of values* $\alpha_j' = \hat{U}_{n-p+j}$, $1 \leqslant j \leqslant q$, *which generally differ from the initial* α_j (cf. (13)).

At this point, we have completed **one step** of the iteration, which to $\alpha \in \mathbb{R}^q$ has associated $\alpha' = \varphi(\alpha) \in \mathbb{R}^q$, where the function φ implicitly described by the above procedure depends only on a, b, $X(n)$ and the accuracy of computations.

The iteration then proceeds as follows. Start with $\alpha(0) = 0 \in \mathbb{R}^q$. Compute successively the vectors $\alpha(k)$ given by $\alpha(k+1) = \varphi[\alpha(k)]$. We shall see in Section 4 that *when X has invertible rational spectrum, and if the number of observations n is large enough, the sequence $\alpha(k)$ converges at exponential speed towards* $\alpha(\infty)$. *moreover the $\hat{W}_m[k]$ computed at step k converge for each fixed m, and for $k \to \infty$, towards the true values $\hat{W}_m = E[W_m|X(n)]$. Finally one has*

$$F_n[a,b,X(n)] = \lim_{k\to+\infty} \sum_{m\leqslant n} |\hat{W}_m(k)|^2.$$

Note also the particular case $q = 0$ where one single step yields the desired value of F_n.

3. Computation of the Information Matrix

3.1. A More General Parameter Space

To study, in Chapter 13, situations where the actual dimensions (p,q) are unknown, we shall consider a centered gaussian ARMA process X with spectral density

$$f_\theta(\lambda) = \frac{\sigma^2}{2\pi}\left|\frac{Q}{P}(e^{-i\lambda})\right|^2$$

where

$$\theta = (P,Q,\sigma), \qquad P(0) = Q(0) = 1,$$

degree $P \leq p$, degree $Q \leq q$, and PQ has all its roots of modulus > 1. *Note that the irreducibility of Q/P is not assumed anymore.* The set of such P, Q is of course an open subset of the euclidean space $\mathbb{R}^{p+q}$, using as coordinates $a_1 \ldots a_p$, $b_1 \ldots b_q$ the $(p + q)$ free coefficients in P, Q. We point out that the canonical spaces Θ used in 1.3 are in fact subsets of the parameter space Θ introduced here, and that their interior points are also interior points of Θ, since Θ is an open subset of $\mathbb{R}^{p+q+1}$.

We want to compute $J(\theta)$ and describe its kernel.

3.2. Computation of the Information Matrix

We introduce *the vector space* $V \simeq \mathbb{R}^{p+q}$ of all pairs (A,B), where A, B are arbitrary polynomials of degree $d°A \leq p$, $d°B \leq q$, with $A(0) = B(0) = 0$, and where $s \in \mathbb{R}$ is arbitrary. Of course $V \times \mathbb{R}$ is the natural tangent space of Θ, and will be trivially identified with $\mathbb{R}^{p+q+1}$.

Fix $\lambda \in [-\pi,\pi]$ and set $z = e^{-i\lambda}$ to write, since

$$f_\theta(\lambda) = \frac{\sigma^2}{2\pi}\left|\frac{Q}{P}(z)\right|^2,$$

$$\log f_\theta(\lambda) = +\log\left(\frac{\sigma^2}{2\pi}\right) + \log Q(z) + \log Q(1/z)$$
$$- \log P(z) - \log P(1/z).$$

Thus for $\theta = (P,Q,\sigma)$ and $v = (A,B,s) \in V \times \mathbb{R}$ we immediately get

$$(16)\qquad \frac{\partial[\log f_\theta(\lambda)]}{\partial\theta}\cdot v = -\frac{A}{P}(z) - \frac{A}{P}\left(\frac{1}{z}\right) + \frac{B}{Q}(z) + \frac{B}{Q}\left(\frac{1}{z}\right) + \frac{2}{\sigma}s$$

and by formula (8)

$$(17)\qquad v^*J(\theta)v = \frac{1}{4\pi}\int \left|\frac{\partial[\log f_\theta(\lambda)]}{\partial\theta}\cdot v\right|^2 d\lambda.$$

The variables $\beta = (P,Q)$ and σ can actually be separated. Indeed write

$$g_\beta(\lambda) = \frac{1}{2\pi}\left|\frac{Q}{P}(e^{-i\lambda})\right|^2$$

so that $f_\theta = \sigma^2 g_\beta$. By Theorem 4.3, Chapter 7, we have

$$\log \sigma^2 = \frac{1}{2\pi}\int_{\mathbb{T}} \log 2\pi f(\lambda)d\lambda$$

which implies

$$(18)\qquad \int_{\mathbb{T}} \log 2\pi g_\beta(\lambda)d\lambda \equiv 0$$

and hence taking the derivative in β

$$(19)\qquad \int_{\mathbb{T}} \frac{\partial}{\partial\beta}[\log f_\theta(\lambda)]d\lambda \equiv \int_{\mathbb{T}} \frac{\partial}{\partial\beta}[\log g_\beta(\lambda)]d\lambda \equiv 0.$$

Since

$$\frac{\partial}{\partial\sigma}[\log f_\theta(\lambda)] = \frac{2}{\sigma}$$

is independent of λ, we get

$$(20)\qquad \int_{\mathbb{T}} \frac{\partial}{\partial\sigma}[\log f_\theta]\frac{\partial}{\partial\beta}[\log f_\theta]d\lambda = 0$$

and thus by (16) and (17), for $v = (A,B,s) \in V \times \mathbb{R}$

$$(21)\qquad v^*J(\theta)v = \frac{1}{4\pi}\left[\int_{\mathbb{T}} |H(z) + H(1/z)|^2 d\lambda\right] + \frac{2}{\sigma^2}s^2$$

where

$$z = e^{-i\lambda},\quad H(z) = \frac{B}{Q}(z) - \frac{A}{P}(z).$$

Since the asymptotic covariance matrix of maximum

likelihood estimators $\hat{\beta}_n$, $\hat{\sigma}_n$ of β, σ is proportional to $J(\theta)^{-1}$, we see that whenever the asymptotic normality Theorem 1.4 holds, then $\hat{\beta}_n$ and $\hat{\sigma}_n$ are *asymptotically uncorrelated.*

Note also that in the same situation the asymptotic standard deviation of $\sqrt{n}(\hat{\sigma}_n - \sigma)$ is, in view of (21), equal to $\sigma/\sqrt{2}$.

3.3. Theorem (kernel of J(θ)). *Consider the gaussian* ARMA *process X parametrized by* $\theta = (P,Q,\sigma) \in \Theta$. *The precise hypotheses are those of Section* 3.1, *and the dimensional parameters of* Θ *are p, q. Consider* $\theta_0 = (P_0,Q_0,\sigma_0) \in \Theta$ *with* $d^\circ P_0 = p_0$ *and* $d^\circ Q_0 = q_0$.

Write $Q_0/P_0 = Q_1/P_1$ *where* Q_1/P_1 *is irreducible, and* $d^\circ Q_1 = q_1$, $d^\circ P_1 = p_1$. *Let* $m = (p - p_1) \wedge (q - q_1)$. *Then the kernel of the information matrix* $J(\theta_0)$ *in the space* $V \times \mathbb{R}^{p+q+1}$ *(introduced in Section* 3.2) *is the space of all vectors* $v \in V \times \mathbb{R}$ *of the form* $v = (UP_1,UQ_1,0)$ *where U is an arbitrary polynomial of degree inferior or equal to m, and* $U(0) = 0$. *In particular the rank of* $J(\theta_0)$ *is equal to* $[p + q + 1 - m]$.

Consequently $J(\theta_0)$ *is invertible if and only if* Q_0/P_0 *is irreducible and at least one of the polynomials* P_0, Q_0 *is of maximal degree (within the constraints imposed by* Θ*).*

Proof. Let $v = (A,B,s) \in V \times \mathbb{R}$ be in the kernel of $J(\theta_0)$. Write

$$H(z) = \frac{B}{Q_0}(z) - \frac{A}{P_0}(z)$$

for $z \in \mathbb{C}$. By (21), we immediately see that we must have $s = 0$ and that the rational fraction $H(z) + H(1/z)$ is equal to zero for $|z| = 1$, which forces the identity $H(z) + H(1/z) \equiv 0$ for all $z \in \mathbb{C}$. But by definition of Θ, the poles of $H(z)$, which are necessarily roots of P_0Q_0, have modulus > 1. Hence if τ were such a pole, we would have

$$\lim_{z \to \tau} |H(z) + H(1/z)| = +\infty$$

which is impossible. Consequently $H(z)$ has no pole. Writing $P_0 = SP_1$, $Q_0 = SQ_1$ we get

$$H = \frac{BP_1 - AQ_1}{SP_1Q_1} ;$$

and hence SP_1Q_1 must be a divisor of $BP_1 - AQ_1$. Since P_1

and Q_1 are coprime, we first get $B = \breve{B}Q_1$, $A = \breve{A}P_1$, and then $\tilde{B} - \tilde{A} = \tilde{H}S$, where $\tilde{A}$, $\tilde{B}$, $\tilde{H}$ are polynomials. At that point we have $H \equiv \tilde{H}$ and hence the identity $H(z) + H(1/z) \equiv 0$ forcfes trivially the polynomial $\tilde{H}$ to be identically 0.

Denote by U the polynomial $\breve{A} \equiv \breve{B}$ to obtain $A = UP_1$, $B = UQ_1$. Since $P_0(0) = Q_0(0) = 1$ we must have $P_1(0) = Q_1(0) \neq 0$, and hence the initial restriction $A(0) = B(0) = 0$ (imposed by the definition of $\mathcal{V}$ as seen in Section 3.2) is equivalent to $U(0) = 0$. The degree condition $d°U \leqslant (p - p_1) \wedge (q - q_1) = m$ comes from the degree restrictions in $\mathcal{V}$.

The sufficiency of these conditions to grant that v lies in the kernel of $J(\theta_0)$ is obvious. And the invertibility statement about $J(\theta_0)$ follows from an elementary analysis of the condition $(p - p_1) \wedge (q - q_1) = 0$. This achieves the proof of Theorem 3.3.

3.4. Remark

It is clear that the proof of Theorem 3.3 has made **no use** of the fact that X is gaussian.

3.5. Explicit Computation of $J(\theta)$

We shall use the method of residuals (cf. [47]). Recall that if $z = e^{-i\lambda}$ and if g is a rational fraction without any zero or pole of modulus 1

$$\int_{-\pi}^{\pi} g(e^{-i\lambda})d\lambda = -\frac{1}{i}\int_{\partial\gamma}\frac{g(z)}{z}dz = 2\pi\sum_{\gamma}\operatorname{res}\left[\frac{g(z)}{z}\right]$$

where $\partial\gamma$ is the boundary of the unit disc γ, oriented in the direction opposed to the standard trigonometric sense, and where the sum Σ_γ is extended to the poles of $g(z)/z$ interior to γ.

Write then $\theta = (\theta_1 \ldots \theta_{p+q+1}) = (a,b,\sigma)$ and $J(\theta) = [J_{k\ell}]_{1 \leqslant k,\ell \leqslant p+q+1}$. Then for instance, for $1 \leqslant k,\ell \leqslant p$ one has

$$4\pi J_{k\ell} = 2\pi\sum_{\gamma}\operatorname{res}\left(\frac{g(z)}{z}\right)$$

where, in view of (16) and (21)

$$g(z) = \left[\frac{z^k}{P(z)} + \frac{z^{-k}}{P(1/z)}\right]\left[\frac{z^\ell}{P(z)} + \frac{z^{-\ell}}{P(1/z)}\right].$$

Since

$$\int_{\partial\gamma} \frac{z^{k+\ell}}{P(z)^2}\frac{1}{z}dz = \int_{\partial\gamma} \frac{z^{-k-\ell}}{P(1/z)^2}\frac{1}{z}\,dz,$$

as shown by the change of variable $z \to 1/z$, and since the first of these integrals is obviously zero by the theorem of residuals, we see that

$$\sum_\gamma \text{res}\left[\frac{g(z)}{z}\right] = \sum_\gamma \text{res}\left[\frac{z^{\ell-k-1} + z^{k-\ell-1}}{P(z)P(1/z)}\right].$$

The poles of modulus < 1 of the last rational fraction are the $1/\tau$ where τ is a root of P, and if τ is a simple root of P, the residual at $1/\tau$ is

$$\left[-\frac{\tau^{k-\ell} + \tau^{\ell-k}}{\tau P'(\tau)P(1/\tau)}\right].$$

Whence the result, when P has no multiple root

$$J_{k\ell} = -\frac{1}{2}\sum_\tau \frac{\tau^{k-\ell-1} + \tau^{\ell-k-1}}{P'(\tau)P(1/\tau)} \quad \text{for } 1 \leqslant k,\ell \leqslant p,$$

where the sums Σ_τ are extended to the roots τ of P. The other terms of $J(\theta)$ are computed in a similar way.

3.6. Comparison of $J(\theta)$ with the Covariances of X

Since

$$r_k = \int e^{ik\lambda} f_\theta(\lambda)d\lambda,$$

the method of residuals gives

$$r_k = \sigma^2 \sum_\gamma \text{res}\, \frac{1}{z^{k+1}} \frac{Q(z)Q(1/z)}{P(z)P(1/z)}.$$

Let us do the computation *in the autoregressive case*, where the result is particularly simple, assuming for the sake of brevity that $P(z)$ has distinct roots. The poles of $1/z^{k+1}P(z)P(1/z)$ interior to γ are the $1/\tau$ with τ root of P, and 0 can only be a pole if $k \geqslant p$. The residual at $1/\tau$ is

$$\left[-\frac{\tau^{k-1}}{P'(\tau)P(1/\tau)}\right]$$

and the residual at 0 is $1/a_p$ when $k = p$ and 0 for $k \geqslant p + 1$.

Whence the formulas, valid for the **autoregressive case,** *and P without multiple root,*

$$r_k = -\sigma^2 \sum_\tau \frac{\tau^{k-1}}{P'(\tau)P(1/\tau)} \qquad \text{for } 0 \leqslant k, \, k \neq p$$

$$r_p = -\sigma^2 \sum_\tau \frac{\tau^{p-1}}{P'(\tau)P(1/\tau)} + \frac{\sigma^2}{a_p} \qquad \text{for } k = p.$$

Thanks to 3.2, one then has

$$J_{k\ell}(\theta) = \frac{r_{k-\ell}}{\sigma^2} \qquad \text{for } 1 \leqslant k, \ell \leqslant p \quad \text{and} \quad (k,\ell) \neq (p,p)$$

$$J_{pp}(\theta) = \frac{r_0}{\sigma^2}.$$

3.7. Numerical Computation of $J(\theta)$

For p, q large enough the numerical resolution of $P(z) = 0$ and $Q(z) = 0$ is longer than the direct numerical computation of the integrals giving $J(\theta)$. However in practical estimation situations, the direct formulas giving $J(\theta)$ in terms of the covariances r_k and the partial autocorrelation a_p (autoregressive case) can be of obvious use.

4. *Convergence of the Backforecasting Algorithm*

4.1. Notations

Let $\tilde{x}_m$, $\tilde{w}_m$, $\tilde{u}_m$ be three numerical sequences. Set

$$u_m = \begin{pmatrix} \tilde{x}_m \\ \tilde{x}_{m-1} \\ \vdots \\ \tilde{x}_{m-p+1} \end{pmatrix} \qquad y_m = \begin{pmatrix} \tilde{x}_m \\ \tilde{x}_{m+1} \\ \vdots \\ \tilde{x}_{m+p-1} \end{pmatrix}$$

(22)

$$w_m = \begin{pmatrix} \tilde{w}_m \\ \tilde{w}_{m-1} \\ \vdots \\ \tilde{w}_{m-q+1} \end{pmatrix} \qquad u_m = \begin{pmatrix} \tilde{u}_m \\ \tilde{u}_{m+1} \\ \vdots \\ \tilde{u}_{m+q-1} \end{pmatrix}.$$

The relations

$$\sum_{k=0}^{p} a_k \breve{x}_{m+k} = \sum_{\ell=0}^{q} b_\ell \breve{u}_{m+\ell} \qquad m \in \mathbb{Z}$$

$$\sum_{k=0}^{p} a_k \breve{x}_{m-k} = \sum_{\ell=0}^{q} b_\ell \breve{w}_{m-\ell} \qquad m \in \mathbb{Z}$$

are obviously equivalent to

$$(23) \qquad y_m - Ay_{m+1} = u_m - Bu_{m+1} \qquad m \in \mathbb{Z}$$

$$(24) \qquad x_m - Ax_{m-1} = w_m - Bw_{m-1} \qquad m \in \mathbb{Z}$$

with A, B square matrices of respective orders p and q, given by

$$A = \begin{bmatrix} -a_1 & \cdots\cdots & & -a_p \\ 1 & 0 & \cdots\cdots & 0 \\ 0 & 1 & \cdots\cdots & 0 \\ \vdots & \vdots & & \\ 0 & 0 & \cdots\cdots 1 & 0 \end{bmatrix}$$

$$B = \begin{bmatrix} -b_1 & \cdots\cdots & & -b_q \\ 1 & 0 & \cdots\cdots & 0 \\ 0 & 1 & \cdots\cdots & 0 \\ \vdots & \vdots & & \\ 0 & 0 & \cdots\cdots 1 & 0 \end{bmatrix}$$

An immediate computation shows that the eigenvalues of A (resp. B) are the inverses of the roots of P (resp. Q) and hence have modulus < 1. When a square matrix M has all its eigenvalues of modulus $< c < 1$, one sees easily, using its Jordan form, that there exists a $K > 0$ such that $\|M^m\| \leqslant Kc^m$ for all $m \geqslant 0$.

Hence, we select $c < 1$ such that all the roots of P, Q are of modulus $> 1/c$. There is then a $K > 0$ such that

$$(25) \qquad \|A^m\| \leqslant Kc^m \text{ and } \|B^m\| \leqslant Kc^m \quad \text{for } m \geqslant 0.$$

From (22) and (23) one gets for $j \geqslant 1$ and $m \in \mathbb{Z}$

$$y_{m-j} = A^j y_m + \sum_{\ell=0}^{j-1} A^\ell [u_{m-j+\ell} - Bu_{m-j+\ell+1}]$$

$$x_{m+j} = A^j x_m + \sum_{\ell=0}^{j-1} A^\ell [w_{m+j-\ell} - Bw_{m+j-\ell-1}]$$

whence, using (25) and letting $K_1 = (K/1-c)(1 + Kc)$

(26) $$|y_{m-j}| \leqslant Kc^j |y_m| + K_1[\sup_{m-j \leqslant k \leqslant m} |u_k|] \quad \text{for } j \geqslant 1,\ m \in \mathbb{Z}$$

(27) $$|x_{m+j}| \leqslant Kc^j |x_m| + K_1[\sup_{m \leqslant k \leqslant m+j} |w_k|] \quad \text{for } j \geqslant 1,\ m \in \mathbb{Z}$$

and by the same argument

(28) $$|u_{m-j}| \leqslant Kc^j |u_m| + K_1[\sup_{m-j \leqslant k \leqslant m} |y_k|] \quad \text{for } j \geqslant 1,\ m \in \mathbb{Z}$$

(29) $$|w_{m+j}| \leqslant Kc^j |w_m| + K_1[\sup_{m \leqslant k \leqslant m+j} |u_k|] \quad \text{for } j \geqslant 1,\ m \in \mathbb{Z}.$$

4.2. Description of the Algorithm

Start with

$$u_{n-p+1} = \alpha \in \mathbb{R}^q \quad \text{and} \quad X = (\tilde{x}_1 \ldots \tilde{x}_n)$$

$$w_{-S} = w \in \mathbb{R}^q \quad \text{and} \quad u_T = u \in \mathbb{R}^q.$$

Consider the following algorithm :

(a) Starting with $u_{n-p+1} = \alpha$ one computes the u_m by (23) with $n - p + 1 \geqslant m \geqslant 1$, m decreasing; set $u_m = 0$ for $-q + 1 \geqslant m$ which, due to the overlap of coordinates for successive u_m, defines all the u_m for $m \leqslant n - p + 1$.

(b) The y_m, $n - p + 1 \geqslant m \geqslant 1$ are directly determined by X. The y_m, $m \leqslant 0$ are computed by (23) with decreasing m. An obvious rewriting of coordinates gives then the x_m for $m \leqslant n$.

(c) Letting $w_{-S} = w$ one computes the w_m, $-S \leqslant m \leqslant n$ by (24) with increasing m. Set $w_m = 0$ for $m \geqslant n + q - 1$, which, due to the overlap of coordinates, determines all the w_m, $m \in \mathbb{Z}$.

(d) By (24), with increasing m, one computes the x_m, $m \geqslant n + 1$.

(e) Setting $u_T = u$, one computes the u_m, $T \geq m \geq n - p + 1$ using (23) with decreasing m. *This provides us with a new value for* u_{n-p+1}. This new value is a function $\psi(\alpha,w,u,X)$ *which is obviously linear* and may be written

$$\psi(\alpha,w,u,X) = \mu\alpha + w\rho + u\tau + \nu X^* \tag{30}$$

where μ, ρ, τ, ν are matrices of order (q,q), $(q,1)$, $(q,1)$, (q,n).

The backforecasting algorithm of Section 2 corresponds to the following particular case: One takes $X = (X_1 \ldots X_n)$, $u = w = 0$, and one computes at the end of step 1

$$\varphi(\alpha) = \psi(\alpha,0,0,X). \tag{31}$$

Setting $\alpha_0 = \alpha$, $\alpha_{k+1} = \varphi(\alpha_k)$, one studies then $\alpha_\infty = \lim_{k\to\infty}\alpha(k)$. The sequence W_m is then estimated by $\hat{W}_m = \tilde{w}_m(\alpha_\infty,0,0,X)$, for $m \geq -S + 1$ and by 0 for $m \leq -S$.

By (28) and (29), one has

$$\varphi(\alpha) - \varphi(\beta) = \mu(\alpha - \beta) \qquad \alpha,\beta \in \mathbb{R}^q.$$

To check that the backforecasting algorithm converges it is sufficient to check that $\|\mu\| < 1$, *which allows us to restrict ourselves to the study of* $\psi(\alpha,0,0,0)$.

4.3. The Contraction Property $\|\mu\| < 1$

Let us then impose $w = u = 0$ and $X = 0$. In phase (a) one has then $y_m = 0$ for $1 \leq m < n - p + 1$ whence by (28)

$$|u_1| \leq Kc^{n-p}|\alpha|.$$

The sliding structure of the coordinates of u_m and the fact that $u_m = 0$ for $m \leq 0$ yields $|u_m| \leq |u_1|$ for $-q + 1 \leq m \leq 1$ and hence

$$|u_m| \leq Kc^{n-p}|\alpha| \qquad \text{for} \quad m \leq 1. \tag{32}$$

In phase (b), (26) with $m = 1$, $j \geq 1$, and (32) give

$$|y_k| \leq K_1Kc^{n-p}|\alpha| \qquad k \leq 0$$

whence, since $|x_m| = |y_{m-p+1}|$ and $x_m = 0$ for $p \leqslant m \leqslant n$,

(33) $$|x_k| \leqslant K_1 K c^{n-p} |\alpha| \qquad k \leqslant n.$$

In phase (c), (29) with $m = -S$, $j \geqslant 1$ and (33) imply $|w_k| \leqslant K_1^2 K c^{n-p} |\alpha|$ for $k \leqslant n$. Since the w_m are zero for $m \geqslant n+1$ one concludes that

(34) $$|w_k| \leqslant K_1^2 K c^{n-p} |\alpha| \qquad \text{for all } k \in \mathbb{Z}\,.$$

In phase (d), (27) and (34) yield

$$|x_k| \leqslant K_1^2 K c^{n-p} |\alpha| \qquad \text{for } k \geqslant n$$

and since $|y_m| = |x_{m+p-1}|$

(35) $$|y_k| \leqslant K_1^3 K c^{n-p} |\alpha| \qquad \text{for } k \geqslant n-p+1.$$

In phase (e), (28) and (35) give, for the new value $\psi(\alpha,0,0,0)$ of u_{n-p+1} the bound

$$|\psi(\alpha,0,0,0)| \leqslant K_1^4 K c^{n-p} |\alpha|$$

and finally $\|\mu\| \leqslant K_1^4 K c^{n-p}$, *which will be* < 1 *provided the number* n *of observations is large enough. The algorithm* $\alpha_{k+1} = \varphi(\alpha_k)$ *converges then for all* S, T *provided* $S \geqslant q+1$ *and* $T \geqslant n-p+1$.

In fact since $K_1^4 K c^{n-p} < 1$ is a number independent of S, T, *one may a priori give oneself an arbitrary sequence* S_k, T_k *of integers such that* $S_k \geqslant q+1$, $T_k \geqslant n-p+1$, *and at each step use the algorithm* φ_k *associated to* S_k, T_k; *the sequence* $\alpha_{k+1} = \varphi_k(\alpha_k)$ *will converge by a classical contraction argument. This corresponds to the practical version of the backforecasting algorithm. Of course the limit* α_∞ *will depend of the particular sequence* S_k, T_k *selected in the computation, but the dependence is quite weak if* $S_k \geqslant S$, $T_k \geqslant T$ *with* S, T *large enough*, a point which we are going to prove in 4.4.

4.4. Study of α_∞

Consider the particular case $S_k \equiv S$, $T_k \equiv T$, to simplify the notations. Assume that S, T have been selected large enough to grant

$$\sum_{m \leqslant -S} |\hat{W}_m|^2 \leqslant \varepsilon$$

(36) $$|\hat{W}_m| \leqslant \varepsilon \quad \text{for } m \leqslant -S$$

$$|\hat{U}_m| \leqslant \varepsilon \quad \text{for } m \geqslant T.$$

This is always possible since $\Sigma_{m \leqslant n} |\hat{W}_m|^2$ and $\Sigma_{m \geqslant 1} |\hat{U}_m|^2$ are finite. Denote by $\hat{x}_m$, $\hat{y}_m$, $\hat{w}_m$, $\hat{u}_m$ and $\overline{w}_m$ the vector sequences associated by (22) to the numerical sequences $\hat{X}_m$, $\hat{W}_m$, $\hat{U}_m$ and W_m, where the W_m can be computed by the preceding algorithm with $u = w = 0$ and the starting point $\alpha = \alpha_\infty$. The W_m are then the approximations of the $\hat{W}_m$ supplied by the backforecasting algorithm, after convergence of the algorithm. Set $\hat{\alpha} = \hat{u}_{n-p+1}$.

One has by definition, with $X = (X_1 ... X_n)$, the relations

$$\hat{\alpha} = \psi(\hat{\alpha}, \hat{w}_{-S}, \hat{u}_T, X)$$

$$\alpha_\infty = \varphi(\alpha_\infty) = \psi(\alpha_\infty, 0, 0, X).$$

Whence by (30)

$$\hat{\alpha} - \alpha_\infty = \mu(\hat{\alpha} - \alpha_\infty) + \hat{w}_{-S}\rho + \hat{u}_T \tau$$

and by (36)

$$|\hat{\alpha} - \alpha_\infty| \leqslant \|\mu\| |\hat{\alpha} - \alpha_\infty| + (\|\rho\| + \|\tau\|)\, \varepsilon$$

whence since $\|\mu\| < 1$

(37) $$|\hat{\alpha} - \alpha_\infty| \leqslant \frac{\|\rho\| + \|\tau\|}{1 - \|\mu\|}\, \varepsilon .$$

For each fixed m with $m \geqslant -S$, the vector w_m obtained by the algorithm (a), (b), and (c) is a **linear** function $\chi_m(\alpha, w, X)$. In particular,

$$\hat{w}_m = \chi_m(\hat{\alpha}, \hat{w}_{-S}, X) \quad \text{and} \quad \overline{w}_m = \chi_m(\alpha_\infty, 0, X)$$

whence

(38) $$\hat{w}_m - \overline{w}_m = \chi_m(\hat{\alpha} - \alpha_\infty, \hat{w}_{-S}, 0).$$

The same computations as in 4.3, applied to $\psi(0,w,0,0)$, $\psi(0,0,u,0)$ and $\chi_m(\alpha,w,0)$ prove *the existence of K_2 independent of S, T such that $\|\rho\|$, $\|\tau\|$, and $\|\chi_m\|$, $m \geqslant -S$ are bounded by K_2 for all S, T large enough.* From (37), (38) and (36) one then gets

$$|\hat{\alpha} - \alpha_\infty| \leqslant K_3\varepsilon \quad \text{and} \quad |\bar{w}_m - \hat{w}_m| \leqslant K_4\varepsilon \text{ for } m \leqslant n,$$

where the constants K_3, K_4 do not depend on S, T. This implies

$$(39) \qquad \left| \sum_{m \leqslant n} |\hat{W}_m|^2 - \sum_{-S \leqslant m \leqslant n} |\bar{W}_m|^2 \right| \leqslant \varepsilon[1 + (n + S)K_4].$$

Since the $\hat{W}_k$ tend to 0 at exponential speed, one can then, in (36), select $\varepsilon = K_5 \exp(-K_6 S)$ where K_5, $K_6 > 0$ are constants.

Inequality (39) shows that $\lim_{S\to+\infty}(F_n - \bar{F}_n) = 0$ where $\bar{F}_n$ is the approximation of F_n given by the backforecasting algorithm.

Bibliographical Hints

The asymptotic results of this chapter will be proved in Chapter 13 at the end of which we sketch the corresponding bibliography. Let us suggest, as further reading, Chapter 6 in *Hannan*. The backforecasting algorithm is described in details by *Box-Jenkins*, but the proof given here for the convergence of this algorithm seems to be new.

Chapter XIII
ASYMPTOTIC MAXIMUM LIKELIHOOD

1. Approximate Log-Likelihood

1.1. Two Useful Approximations

Let X be a stationary, centered, nonzero gaussian process, with spectral density f. Let μ_n be the law of $X_1 ... X_n$, h_n the density of μ_n on $\mathbb{R}^n$ and $\mathfrak{L}_n(f, X_1 ... X_n) = \log h_n(X_1 ... X_n)$ the log-likelihood of X. We have seen (Chapter 12, Section 1.2) that

$$(1) \qquad -2\mathfrak{L}_n(f, X_1 ... X_n) = n \log 2\pi + \log \det T_n(2\pi f) + X(n)^*[T_n(2\pi f)]^{-1}X(n)$$

where T_n is the Toeplitz matrix and $X(n)^* = (X_1 ... X_n)$.

We shall have to study the behaviour of $\mathfrak{L}_n$ when the spectral density of X is different of f. When X is gaussian as above, and has an arbitrary spectral density, this leads us to set, for every even positive bounded function g: $\mathbb{T} \to \mathbb{R}^+$,

$$(2) \qquad L_n(g, X) = -\frac{1}{2}\left[n \log 2\pi + \log \det T_n(2\pi g) + X(n)^*[T_n(2\pi g)]^{-1}X(n)\right].$$

In Chapter 12 we have, for X regular, approached $(1/n)\log \det T_n(2\pi f)$ by its limit $\log \sigma^2$ which is the logarithm of the one step prediction error and hence is equal to

$$\frac{1}{2\pi}\int_{\mathbb{T}} \log(2\pi f)d\lambda.$$

This is equivalent to the replacement of L_n by $\tilde{L}_n$ where

$$\tilde{L}_n(g,X) = -\frac{1}{2}\Big[n \log 2\pi + \frac{n}{2\pi}\int_{\mathbb{T}} \log(2\pi f)d\lambda \tag{3}$$
$$+ X(n)^*[T_n(2\pi g)]^{-1}X(n)\Big].$$

On the other hand for n large, $(1/n)T_n$ is "almost" a homomorphism (cf. Chapter 10, Section 5.1) which suggests the replacement of $[T_n(2\pi g)]^{-1}$ by $T_n(1/2\pi g)$ and introduces **the Whittle approximation** $\bar{L}_n$,

$$\bar{L}_n(g,X) = -\frac{1}{2}\Big[n \log 2\pi + \frac{n}{2\pi}\int_{\mathbb{T}} \log(2\pi g)d\lambda \tag{4}$$
$$+ X(n)^*T_n\left(\frac{1}{2\pi g}\right)X(n)\Big]$$

which (cf. Chapter 10), using the periodogram I_n of X may be rewritten

$$\bar{L}_n(g,X) = -\frac{n}{2}\left[\log 2\pi + \frac{1}{2\pi}\int_{\mathbb{T}} \log(2\pi g)d\lambda + I_n\left(\frac{1}{2\pi g}\right)\right]. \tag{5}$$

The Whittle approximation is a very efficient theoretical tool while the approximation $\tilde{L}_n$ is better suited to numerical computations in the ARMA case, as we have seen in Chapter 12.

For any suitable function φ, with Fourier coefficients $\hat{\varphi}_k$, we set

$$\alpha(\varphi) = \sum_k (1 + |k|)|\hat{\varphi}_k|$$

as in Chapter 10, Sections 3 and 5. We shall have to use hypotheses of the type $\alpha(\log f) < +\infty$. *Note that if* $\rho = \alpha(\log f)$, *then - see Section 5 below -* $\alpha(f)$ *and* $\alpha(1/f)$ *are bounded by* e^{ρ}, *the range of* f *is included in* $[e^{-\rho},e^{\rho}]$, *and* f *has a continuous derivative bounded by* e^{ρ}.

1.2. Lemma. *Let X be stationary, centered, gaussian, with spectral density f and global law P_f, and let g: $\mathbb{T} \to \mathbb{R}^+$ be such that $\alpha(\log g) \leqslant \rho$, $\alpha(\log f) \leqslant \rho$ with ρ finite. Then there is a number R depending only on ρ such that*

$$(6)\qquad |L_n(g,X) - \hat{L}_n(g,X)| \leqslant R \quad \textit{for all } n \geqslant 1$$

$$(7)\qquad E_{P_f}[|L_n(g,X) - \bar{L}_n(g, X)|^2] \leqslant R \quad \textit{for all } n \geqslant 1.$$

Proof. This result is a consequence of the technical lemmas 4.6 and 4.7 below. Let us add a useful corollary: *under the hypothesis* $\alpha(\log f) \leqslant \rho$, $\alpha(\log g) \leqslant \rho$ one has surely

$$(8)\qquad \lim_{n\to\infty} \left|\frac{1}{n} L_n(g,X) - \frac{1}{n}\hat{L}_n(g,X)\right| = 0 \ \text{(uniformly in } g)$$

and P_f-almost surely

$$(9)\qquad \lim_{n\to\infty} \left|\frac{1}{n} L_n(g,X) - \frac{1}{n}\bar{L}_n(g,X)\right| = 0.$$

Indeed (8) follows immediately from (6). On the other hand (7) implies

$$P_f\left[\left|\frac{1}{n} L_n(g,X) - \frac{1}{n}\bar{L}_n(g,X)\right| \geqslant \frac{1}{n^{1/4}}\right] \leqslant \frac{R}{n^{3/2}}$$

which, by Borel-Cantelli's lemma grants the existence of a finite random integer $N(\omega)$ such that

$$\left|\frac{1}{n} L_n - \frac{1}{n}\bar{L}_n\right| \leqslant \frac{1}{n^{1/4}} \qquad \text{for } n \geqslant N(\omega).$$

2. Kullback Information

2.1. General Case

Let μ, ν be two probabilities on $\mathbb{R}^k$. When $d\nu/d\mu$ exists, **the Kullback information** $K(\mu,\nu)$ **of** ν **with respect to** μ is defined by

$$(10)\qquad K(\mu,\nu) = \int_{\mathbb{R}^k} \left[-\log \frac{d\nu}{d\mu}\right] d\mu.$$

Since $-\log x \geqslant 1 - n$, one has $K(\mu,\nu) \geqslant 0$ and $K(\mu,\nu)$ can only be zero if $\nu = \mu$. When $d\nu/d\mu$ does not exist, one sets $K(\mu,\nu) = +\infty$.

The "distance" defined by K on the space of probability measures plays an essential part in asymptotic statistics (cf. [15], Vol. 2, Chapter 3).

If μ and ν are two probabilities on the infinite product $\mathbb{R}^{\mathbb{N}}$, with projections μ_n and ν_n on $\mathbb{R}^n$, we shall define **the asymptotic information** $K(\mu,\nu)$ **of** ν **with respect to** μ **by**

$$K(\mu,\nu) = \lim_{n\to\infty} \frac{1}{n} K(\mu_n,\nu_n)$$

if this limits exists. The normalizing factor is suggested by the case where μ and ν are product measures $\mu = \mu_1^{\otimes\mathbb{N}}$, $\nu = \nu_1^{\otimes\mathbb{N}}$, a very simple and common situation, where $K(\mu,\nu) = K(\mu_1,\nu_1)$.

2.2. The Gaussian Case

Let X be a stationary, centered, nonzero gaussian process, with arbitrary spectral density g. For each given g, X is defined on the probability space (Ω,P_g) and we denote by $\mu_n(g)$ the distribution of $X_1 \dots X_n$. The preceding formula naturally defines **the information of** P_g **with respect to** P_f **by**

$$K(P_f\,,\,P_g) = \lim_{n\to\infty} \frac{1}{n} K[\mu_n(f),\, \mu_n(g)] \tag{11}$$

if the limits exists. The Lebesgue density of $\mu_n(g)$ on $\mathbb{R}^n$ is $\exp L_n(g,X)$ at the point

$$X(n) = \begin{bmatrix} X_1 \\ \cdot \\ \cdot \\ \cdot \\ X_n \end{bmatrix}.$$

Formulas (10) and (11) hence imply

$$K(P_f,\,P_g) = \lim_{n\to\infty} \frac{1}{n} E_{P_f}[L_n(f,X) - L_n(g,X)] \tag{12}$$

if the limit exists. We are going to compute K for $\alpha(\log f)$ and $\alpha(\log g)$ finite.

2.3. Theorem. *Let f,g be two spectral densities on $\mathbb{T}$ such that $\alpha(\log f)$ and $\alpha(\log g)$ be finite. Let P_f and P_g be the global distributions of two stationary gaussian processes, centered, with* *spectral densities* f,g. *Then the asymptotic information* $K(P_f, P_g)$ *exists and is given by*

$$(13) \qquad K(P_f, P_g) = \frac{1}{4\pi} \int_{\mathbb{T}} \left[-\log \frac{f}{g} - 1 + \frac{f}{g} \right] d\lambda.$$

Moreover, P_f-a.s., *one has*

$$(14) \qquad K(P_f, P_g) = \lim_{n\to\infty} \frac{1}{n} [L_n(f,X) - L_n(g,X)]$$

with uniform convergence on $\{g \mid \alpha(\log g) \leq \rho\}$, *where* ρ *is fixed but arbitrary. Finally properties* (12), (13) *and* (14) *remain true for* $\bar{L}_n$ *and* $\tilde{L}_n$.

Proof. Set $K_n = (1/n)[L_n(f,X) - L_n(g,X)]$ and define similarly $\bar{K}_n$, $\tilde{K}_n$. For φ even and bounded, the sequence $I_n(\varphi)$ converges P_f-a.s. toward $I(\varphi) = \int_{\mathbb{T}} \varphi f \, d\lambda$ and $E_{P_f}[I_n(\varphi)]$ converges toward $I(\varphi)$ (cf. Chapter 10, Section 3.1). Since

$$\bar{K}_n = -\frac{1}{4\pi} \left[\int_{\mathbb{T}} \log \left(\frac{f}{g} \right) d\lambda + I_n\left(\frac{1}{f} - \frac{1}{g} \right) \right]$$

one obtains for each g the P_f-a.s. convergence of $\bar{K}_n$ toward $K(P_f, P_g)$ given by (13), and the convergence of $E_{P_f}(\bar{K}_n)$ toward $K(P_f, P_g)$.

For $\alpha(\log f) \leq \rho$, $\alpha(\log g) \leq \rho$, the functions f, $1/f$, g, $1/g$, $|df/d\lambda|$, $|dg/d\lambda|$ remain bounded by e^ρ. Hence the set $\Theta = \{g \mid \alpha(\log g) \leq \rho\}$ is compact in $C(\mathbb{T})$, $K(P_f, P_g)$ given by (13) is continuous in $(f,g) \in \Theta \times \Theta$, and hence uniformly continuous in $f,g \in \Theta$. Since by definition

$$\left| I_n\left(\frac{1}{h} - \frac{1}{g} \right) \right| \leq \left\| \frac{1}{h} - \frac{1}{g} \right\|_\infty I_n(1)$$

and since $I_n(\lambda)$ converges P_f-a.s. towards $I(1) = \int_{\mathbb{T}} f \, d\lambda$, one sees that, P_f-a.s., the sequence $\bar{K}_n(f,g)$ is an equicontinuous sequence of functions of $g \in \Theta$. An equicontinuous sequence of functions which converges at a dense countable family of points of the compact Θ towards a uniformly continuous limit, must converge at all points and uniformly. Hence P_f-a.s., the sequence $\bar{K}_n(f,g)$ converges, uniformly in $g \in \Theta$, towards $K(P_f, P_g)$ given by (13).

Lemma 1.2 and its consequences (8) and (9) prove that $E_{P_f}(K_n)$, $E_{P_f}(\tilde{K}_n)$ have the same limits as $E_{P_f}(\bar{K}_n)$, and that K_n,

$\breve{K}_n$ have P_f-a.s. the same limit as $\overline{K}_n$.

The P_f-almost sure uniformity for the convergence of K_n towards K follows from the fact that $(1/n)L_n(g,X)$ is P_f-a.s. an equicontinuous sequence in $g \in \Theta$. This equicontinuity is proved as for L_n after having noticed that

$$\|[T_n(g)]^{-1} - [T_n(h)]^{-1}\| \leqslant 2c^2\|T_n(g) - T_n(h)\|$$

since

$$T_n(g) \geqslant \frac{1}{c}I, \quad T_n(h) \geqslant \frac{1}{c}I \text{ and } \|T_n(g) - T_n(h)\| \leqslant \frac{1}{2c}.$$

The P_f-a.s. uniformity in $g \in \Theta$ for the convergence of $\breve{K}_n$ follows then trivially from (8).

2.4. Remark. Since $-\log x - 1 + x \geqslant 0$ and can only be zero for $x = 1$, expression (13) shows that for f, g continuous, one has $K(P_f, P_g) > 0$ unless $f \equiv g$ in which case $K(P_f, P_g) = 0$.

3. Convergence of Maximum Likelihood Estimators

3.1. Construction of the Estimators

Let X be a *stationary gaussian* centered process, whose spectral density f_θ depends on a parameter $\theta \in \Theta$ where Θ *is a topological compact space.* Denote by P_θ instead of P_{f_θ} the global distribution of X, and by $L_n(\theta)$, $\breve{L}_n(\theta)$, $\overline{L}_n(\theta)$ the exact and approximate log-likelihoods $L_n(f_\theta,X)$, $\breve{L}_n(f_\theta,X)$, $\overline{L}_n(f_\theta,X)$ defined by (2), (3) and (4).

Define exact and approximate maximum likelihood estimators θ_n, $\breve{\theta}_n$, $\overline{\theta}_n$ associated to L_n, $\breve{L}_n$, $\overline{L}_n$ by

$$L_n(\theta_n) = \sup_{\theta \in \Theta} L_n(\theta)$$

and analogous formulas for $\breve{\theta}_n$, $\overline{\theta}_n$. For n and $X(n)$ given, *this does not assume the uniqueness of* θ_n, $\breve{\theta}_n$, $\overline{\theta}_n$. *The existence of such estimators is forced* by the compactness of Θ, provided L_n, $\breve{L}_n$, $\overline{L}_n$ are continuous in θ, which will be implied by the following hypothesis:

$$\text{(H)}\quad \begin{cases} H_1\text{: } \alpha(\log f_\theta) \text{ remains bounded for } \theta \in \Theta \\ H_2\text{: for each } \lambda \in \mathbb{T},\ f_\theta(\lambda) \text{ is continuous in } \theta \in \Theta. \end{cases}$$

Hypothesis (H) implies that $\|f_\theta\|_\infty$, $\|1/f_\theta\|_\infty$, $\|df_\theta/d\lambda\|_\infty$ remain bounded for $\theta \in \Theta$, and implies the continuity of $\theta \to f_\theta$ as a map from Θ into $C(\mathbb{T})$. In particular the function $K(\theta,\theta') = K(P_\theta,P_\theta)$, given by (13) is continuous in $(\theta,\theta') \in \Theta \times \Theta$.

Note that we **do not** assume here the injectivity of $\theta \to f_\theta$. In fact we shall say that $\theta \in \Theta$ *is an injective point* of Θ if the set

$$N(\theta) = \{\theta' \in \Theta \mid f_{\theta'} \equiv f_\theta\}$$

is reduced to $\{\theta\}$.

3.2. Theorem. *Assume hypothesis* (H), *with X gaussian and Θ compact. Let θ_n be the maximum likelihood estimator. Then for each $\theta \in \Theta$ the sequence f_{θ_n} converges P_θ-a.s. toward f_θ in $C(\mathbb{T})$. Moreover, P_θ-a.s. the distance between θ_n and the set $N(\theta)$ defined in* 3.1 *converges to zero. In particular if θ is an injective point of Θ, then $\lim_{n\to\infty}\theta_n = \theta$, P_θ-a.s.*

The same properties hold for the approximated maximum likelihood estimators $\overline{\theta}_n$ and $\hat{\theta}_n$.

Proof. Let f_{θ_0} be the true spectral density of X. Let $\ell_n : \Theta \to \mathbb{R}$ be a *deterministic* sequence of continuous functions such that when $n \to \infty$, $|\ell_n(\theta_0) - \ell_n(\theta)|$ converges uniformly in $\theta \in \Theta$, towards $K(\theta_0,\theta)$. Let $v_n \in \Theta$ be such that

$$\sup_{\theta\in\Theta} \ell_n(\theta) = \ell_n(v_n).$$

Then all the limit points of (v_n) belong to $N(\theta_0)$.

Indeed let v_∞ be the limit of an arbitrary subsequence v_{n_k}. One has

$$K(\theta_0,v_\infty) = \lim_{k\to\infty} K(\theta_0,v_{n_k}) = \lim_{k\to\infty} [\ell_{n_k}(\theta_0) - \ell_{n_k}(v_{n_k})]$$

and by definition of the v_n, $\ell_n(\theta_0) \leqslant \ell_n(v_n)$ so that $0 \leqslant K(\theta_\infty,v_\infty) \leqslant 0$ whence $f_{\theta_0} = f_{v_\infty}$ by 2.4, and finally $v_\infty \in N(\theta_0)$

by Definition 3.1.

By Theorem 2.3 this argument can be applied P_{θ_0}-a.s. to the three sequences of functions ℓ_n, $\breve{\ell}_n$, $\bar{\ell}_n$ defined by $\ell_n(\theta) = (1/n)L_n(\theta)$ etc. ... which yields the desired result.

3.3. Remark. Fix an *arbitrary bounded* numerical sequence $\varepsilon_n > 0$ and let θ_n' be an estimator such that

$$\sup_{\theta\in\Theta} L_n(\theta) \geq L_n(\theta_n') \geq [\sup_{\theta\in\Theta} L_n(\theta)] - \varepsilon_n$$

which corresponds to the practical situations where *one maximizes* L_n *approximately.* The same argument shows that Theorem 3.2 holds for θ_n'. A similar conclusion holds for $\breve{L}_n$, $\bar{L}_n$.

4. Asymptotic Normality and Efficiency

4.1. Efficiency and Information Matrix

Let $Y \in \mathbb{R}^k$ be a random vector whose distribution has a density Φ_θ (with respect to Lebesgue measure) depending on the parameter $\theta \in \Theta$ where Θ is open in $\mathbb{R}^k$. The **Fischer information matrix** of Φ_θ is defined by

$$J(\theta) = E_\theta\left[\frac{\partial}{\partial\theta}\log\Phi_\theta(Y)\right]\left[\frac{\partial}{\partial\theta}\log\Phi_\theta(Y)\right]^* = \int_{\mathbb{R}^n}\frac{1}{\Phi_\theta}\Phi_\theta'\Phi_\theta'^*\,dy.$$

When $J(\theta)$ is well defined and invertible the **CramerRao inequality** grants that for any estimator $\hat{\theta}$ of θ based on the observation Y and such that $E_\theta(\hat{\theta}) = \theta$, the covariance matrix Γ_θ of $\hat{\theta}$ verifies $\Gamma_\theta \geq J(\theta)^{-1}$, and $\hat{\theta}$ is said to be **efficient** if $\Gamma_\theta = J(\theta)^{-1}$.

Let now X be a random process with discrete time, whose global distribution P_θ depends on the parameter θ with Θ as above. Assume that the distribution of

$$X(n) = \begin{bmatrix} X_1 \\ \cdot \\ \cdot \\ \cdot \\ X_n \end{bmatrix}$$

has a density on $\mathbb{R}^n$, denoted by exp $L_n(\theta,X)$ at point $X(n)$. The information matrix $J_n(\theta)$ relative to the observation $Y = X(n)$ becomes here

$$J_n(\theta) = E_{P_\theta}\left[\frac{\partial}{\partial\theta} L_n(\theta,X)\right]\left[\frac{\partial}{\partial\theta} L_n(\theta,X)\right]^* .$$

If $\hat{\theta}_n$ is an arbitrary estimator of θ, based on $X_1 \dots X_n$, and such that $E_{P_\theta}(\hat{\theta}_n) = \theta$ one has then (Cramer-Rao)

$$\text{cov}[\sqrt{n}(\hat{\theta}_n - \theta)] \geq nJ_n(\theta)^{-1} = \left[\frac{1}{n} J_n(\theta)\right]^{-1} .$$

This suggests the definition of the *asymptotic information matrix for the family of processes* (X,P_θ) *by*

$$J(\theta) = \lim_{n\to\infty} \frac{1}{n} J_n(\theta)$$

(15)

$$= \lim_{n\to\infty} \frac{1}{n} E_{P_\theta}\left[\frac{\partial L_n}{\partial\theta}(\theta,X)\right]\left[\frac{\partial L_n}{\partial\theta}(\theta,X)\right]^*$$

if the limit exists. If moreover $J(\theta)$ *is invertible*, we shall have *for every sequence* $\hat{\theta}_n$ *of unbiased estimators*

$$\varliminf_{n\to\infty} \text{cov}[\sqrt{n}(\hat{\theta}_n - \theta)] \geq J(\theta)^{-1}$$

and we shall say that a sequence of estimators $\hat{\theta}_n$ is **asymptotically efficient** if

$$\lim_{n\to\infty} \text{cov}[\sqrt{n}(\hat{\theta}_n - \theta)]$$

is equal to $J(\theta)^{-1}$.

4.2. Hypotheses

Consider a stationary centered gaussian process, whose spectral density f_θ depends on the parameter $\theta \in \Theta$ where Θ *is a compact subset of* $\mathbb{R}^k$. We assume that

NA1: the map $\theta \to f_\theta(\lambda)$ is for each $\lambda \in \mathbb{T}$, the restriction to Θ of a function of class 2 on some open subset of $\mathbb{R}^k$

NA2: the numbers $\alpha(\log f_\theta)$, $\alpha(\partial_i f_\theta)$, $\alpha(\partial^2_{ij} f_\theta)$ remain bounded when θ varies in Θ - with the notations $\partial_i = \partial/\partial\theta_i$, $\partial^2_{ij} = \partial^2/\partial\theta_i\partial\theta_j$.

For instance the hypotheses NA1, NA2 *are certainly satisfies if* $(\theta,\lambda) \to f_\theta(\lambda)$ *is a function of class* 3 *on a neighborhood of* $\Theta \times [-\pi,\pi]$ *and if* $f_\theta(\lambda) > 0$ *for all* (θ,λ) *in* $\Theta \times [-\pi,\pi]$.

Let us state the two essential results of this chapter.

4.3. Theorem. *Let* X *be a centered stationary gaussian process parametrized by* $\theta \in \Theta$ *compact subset of* $\mathbb{R}^k$. *Assume that the hypotheses* NA1, NA2 *hold. Then the asymptotic information matrix* $J(\theta)$ *defined by* (15) *exists and is equal to*

$$(16) \qquad J(\theta) = \frac{1}{4\pi}\int_{\mathbb{T}} \left[\frac{\partial \log f_\theta}{\partial \theta}\right]\left[\frac{\partial \log f_\theta}{\partial \theta}\right]^* d\lambda, \qquad \theta \in \Theta.$$

Moreover if $L_n(\theta,X) = L_n(f_\theta,X)$ *is the log-likelihood of* X *one has for each* $\theta \in \Theta$

$$(17) \qquad J(\theta) = \lim_{n\to\infty} \left[-\frac{1}{n}\frac{\partial^2}{\partial\theta^2} L_n(\theta,X)\right], \qquad P_\theta\text{-a.s.}$$

Finally, the relations (15), (16) and (17) *remain true when* L_n *is replaced by* $\tilde{L}_n$ *or* $\bar{L}_n$.

4.4. Theorem. *Let* X *be a centered stationary gaussian process parametrized by* $\theta \in \Theta$ *compact subset of* $\mathbb{R}^k$. *Assume that* NA1, NA2 *are satisfied. Then for every point* θ **interior** *to* Θ *and* **injective** *in* Θ (cf. 3.1) *the exact and approximate maximum likelihood estimators* θ_n, $\tilde{\theta}_n$, $\bar{\theta}_n$ *converge* P_θ-a.s. *toward* θ. *Moreover if the information matrix* $J(\theta)$ *is invertible, the distributions of* $\sqrt{n}(\theta_n - \theta)$, $\sqrt{n}(\tilde{\theta}_n - \theta)$, $\sqrt{n}(\bar{\theta}_n - \theta)$ *converge tightly as* $n \to \infty$ *to the gaussian distribution* $N(0,J(\theta)^{-1})$. *In particular the estimators* θ_n, $\tilde{\theta}_n$ *and* $\bar{\theta}_n$ *are asymptotically efficient.*

Proofs. Theorems 4.3 and 4.4 will be proved in 4.8 and 4.10 below after a few technical lemmas.

Remark. If one only maximizes L_n, $\tilde{L}_n$, and $\bar{L}_n$ within an ε_n-approximation as in 3.3, which corresponds to the concrete computations, the estimators thus obtained still satisfy 4.4, *provided* $\varepsilon_n \to 0$ *as* $n \to \infty$.

4.5. Lemma. *Let* $q_1 \ldots q_r$ *be integers of arbitrary signs. Let* $g_1 \ldots g_r$ *be functions of* $\lambda \in \mathbb{T}$, *depending on the parameter* $\theta \in \Theta$, *and of class* 2 *in* θ. *Assume that, when* θ *varies in* Θ, *the* $\alpha(g_k)$, $\alpha(\partial_i g_k)$, *and* $\alpha(\partial^2_{ij} g_k)$ *remain bounded by a finite number* R, *and* *that for all k such that* $q_k < 0$, *the* $\alpha(\log g_k)$ *remain bounded by* R. *Let* T_n *be the Toeplitz operators. Then the matrices*

$$M_n(\theta) = (T_n g_1)^{q_1} \dots (T_n g_k)^{q_r} - T_n(g_1^{q_1} \dots g_r^{q_r})$$

as well as their derivatives (in θ*) of order* $\leqslant 2$ *remain bounded in block-norm* (cf. *Chapter* 10, *Section* 5, (17)) *when* (n,θ) *varies in* $\mathbb{N} \times \Theta$.

Proof. For g such that $\alpha(\log g) < \infty$ set $h = \log g$, whence for $u \in \mathbb{R}$

$$\alpha(g^u) = \alpha(e^{uh}) \leqslant \sum_{k \geqslant 0} \alpha\left[\frac{u^k h^k}{k!}\right] \leqslant \sum_{k \geqslant 0} \frac{|u|^k[\alpha(h)]^k}{k!} \tag{18}$$
$$= e^{|u|\alpha(\log g)}$$

since $\alpha(\varphi\psi) \leqslant \alpha(\varphi)\alpha(\psi)$. In particular $\alpha(g)$ and $\alpha(1/g)$, and hence a fortiori $\|g\|_\infty$, $\|dg/d\lambda\|_\infty$ are bounded by $e^{\alpha(\log g)}$. Moreover one has

$$T_n(e^{uh}) - e^{uT_n(h)} = \sum_{k \geqslant 0} \frac{u^k}{k!}(T_n(h^k) - [T_n(h)]^k)$$

whence by Proposition 5.1, Chapter 10, the block-norm bound, valid for all $n \in \mathbb{N}$,

$$b[T_n(e^{uh}) - e^{uT_n(h)}] \leqslant \sum_{k \geqslant 0} \frac{u^k}{k!} k\alpha(h)^k \leqslant |u|\alpha(h)e^{|u|\alpha(h)}$$

which is equivalent to

$$b[T_n(g^u) - [T_n(g)]^u] \leqslant |u|\alpha(\log g)e^{|u|\alpha(\log g)}. \tag{19}$$

In particular one has

$$b[T_n(g^{-1}) - [T_n(g)]^{-1}] \leqslant \alpha(\log g)e^{\alpha(\log g)}. \tag{20}$$

A fortiori one has the bound in line-norm (cf. (17) and (18) in Chapter 10) for $u \in \mathbb{R}$

$$\ell\{[T_n(g)]^u\} \leqslant \ell[T_n(g^u)] + |u|\alpha(\log g)e^{|u|\alpha(\log g)} \tag{21}$$
$$\leqslant [1 + |u|\alpha(\log g)]e^{|u|\alpha(\log g)}$$

since $\ell[T_n(\varphi)] \leqslant \alpha(\varphi)$.

To study the $M_n(\theta)$, the q_j being integers, one can always, through the artificial increase of the number of functions from r to $(|q_1| + |q_2| + ... + |q_r|)$, *reduce the problem to the case where all the q_j are equal to* ± 1. Formulas (20) and (21), Proposition 5.1 of Chapter 10, and the elementary inequalities (18) of Chapter 10 prove then immediately by induction on r, that the $M_n(\theta)$ remain bounded in block-norm when n, θ vary, and that this bound only depends on R.

Let D be one of the operators ∂_i or ∂^2_{ij}. The image by D of a **noncommutative monomial** $M = A_1^{q_1} ... A_r^{q_r}$, with integers $q_1 ... q_r$ having arbitrary signs, is a universal function Φ of $A_1 ... A_r$, $\partial_i A_1 ... \partial_i A_r$, $\partial_j A_1 ... \partial_j A_r$, $\partial^2_{ij} A_1 ... \partial^2_{ij} A_r$, where Φ is a finite sum of **noncommutative monomials** with integral coefficients. Write T instead of T_n, and notice that T commutes with the ∂_i, ∂^2_{ij}. Hence if $A_k = Tg_k$, $1 \leqslant k \leqslant r$, one has

$$DM = \Phi(Tg_1 ...; T\partial_i g_1 ...; T\partial_j g_1 ...; T\partial^2_{ij} g_1 ...)$$

When ψ is a monomial in S variables we have seen above that $\psi \circ T^{\otimes S} - T \circ \psi$ is bounded in block-norm provided the argument $\varphi_1 ... \varphi_S$ remains constrained by bounds on $\alpha(\log \varphi_j)$ or $\alpha(\varphi_j)$ according to the signs of the corresponding exponents. Since Φ is a finite sum of monomials, one sees that $(DM - Th)$ remains bounded in block-norm, where

$$h = \Phi(g_1 ...; \partial_i g_1 ...; \partial_j g_1 ...; \partial^2_{ij} g_1 ...).$$

Since Φ is universal, one has the identity

$$h \equiv D(g_1^{q_1} g_2^{q_2} ... g_r^{q_r}).$$

Since $DT = TD$ we see that $T(h) = DT(g_1^{q_1} ... g_r^{q_r})$ which achieves the proof.

4.6. Lemma. *Let X be stationary gaussian as in* 4.2. *Set* $\psi_n(\theta) = \log \det T_n(f_\theta)$ *and*

$$\varphi_n(\theta) = \frac{n}{2\pi} \int \log f_\theta \, d\lambda.$$

Then the functions $(\psi_n - \varphi_n)$ *and their derivatives of order* $\leqslant 1$ *in* θ *remain bounded for* $\theta \in \Theta$, *by a constant independent of* $n \in \mathbb{N}$. *In particular the same result is true for*

$$L_n(\theta,X) - \tilde{L}_n(\theta,X) = -\frac{1}{2}[\psi_n(\theta) - \varphi_n(\theta)].$$

Proof. Let $M(u)$ be a square invertible matrix depending smoothly on $u \in \mathbb{R}$. One then has

$$\text{(22)} \qquad \frac{d}{du} \log \det M(u) = \text{tr}[M(u)^{-1}M'(u)] \qquad u \in \mathbb{R}$$

Indeed, letting $\tilde{M}(v) = M(u)^{-1}M(u + v)$, it is sufficient to prove (22) for $u = 0$, $M(0) = I$, one then has $M(u) = I + uM'(0) + uC(u)$ where $C(u) \to 0$ as $u \to 0$. Whence

$$\log \det M(u) = \log \det[I + uM'(0)] + \log \det[I + uD(u)]$$

where $D(u) = [I + uM'(0)]^{-1}C(u)$ tends to zero as $u \to 0$. Let τ_k be the eigenvalues of $M'(0)$; then

$$\det[I + uM'(0)] = \prod_k (1 + u\tau_k) = 1 + u \;\text{tr}[M'(0)] + 0(u)$$

and similarly $\det[I + uD(u)] = 1 + 0(u)$, whence $\log \det M(u) = u \;\text{tr}[M'(0)] + 0(u)$ which proves (22) in the case $u = 0$, $M(0) = I$ and hence in the general case.

Apply (22) to $M(u) = T_n[f_\theta^u]$ to get

$$\log \det T_n(\theta) = \int_0^1 \frac{d}{du}[\log \det M(u)]du$$

$$= \int_0^1 \text{tr}[T_n(f_\theta^u)^{-1}T_n(f_\theta^u \log f_\theta)]du.$$

Since

$$\text{tr}\; T_n(h) = \frac{n}{2\pi} \int_{\mathbb{T}} h \, d\lambda,$$

we end up with

$$\psi_n(\theta) - \varphi_n(\theta) = \int_0^1 \text{tr}\; Q(n,\theta,u)du$$

where

$$Q(n,\theta,u) = [T_n(f_\theta^u)]^{-1}T_n(f_\theta^u \log f_\theta) - T_n(\log f_\theta).$$

But $\alpha(\log f_\theta^u) = |u|\alpha(\log f_\theta)$ and $\alpha(gh) \leq \alpha(g)\alpha(h)$, so that hypothesis 4.2 and Lema 4.5 prove that $Q(n,\theta,u)$ and its derivatives (in θ) $\partial_i Q$, $\partial^2_{ij} Q$ remain bounded in block-norm for

$n \in \mathbb{N}$, $\theta \in \Theta$, $u \in [0,1]$. Since $\mathrm{tr}(A)$ is bounded by the block-norm of A, and since $\partial_i, \partial^2_{ij}$ commute with $\int_0^1$ under hypothesis 4.2, the lemma is proved.

Note that to check the partial conclusion "$|\psi_n - \varphi_n|$ is bounded", the hypothesis "$\alpha(\log f_\theta)$ is bounded" was sufficient, which proves assertion (6) as stated earlier.

4.7. Lemma. *Let X be gaussian, stationary, as in 4.2. Consider the random variables $Z_{n,\theta} = [L_n(\theta,X) - \bar{L}_n(\theta,X)]$. Then for every fixed $\theta_0 \in \Theta$, every fixed integer $r \geqslant 1$, the random variables $Z_n = \sup_{\theta\in\Theta}|Z_{n,\theta}|$ remain bounded in norm in $L^r(\Omega, P_{\theta_0})$ when n varies in $\mathbb{N}$.*

Proof. We give a detailed proof for $r = 2$, the general case being handled by a similar argument. In view of 4.6 and formulas (2) and (4) it is sufficient to study the variables $Y_{n,\theta} = X(n)^*A_{n,\theta}X(n)$ where

$$A_{n,\theta} = [T_n(f_\theta)]^{-1} - T_n\left(\frac{1}{f_\theta}\right).$$

Hypothesis 4.2 and Lemma 4.6 prove that $A_{n,\theta}$, $\partial_i A_{n,\theta}$, $\partial^2_{ij}A_{n,\theta}$ remain bounded in block-norm by a finite R when n,θ vary. Since $DY_{n,\theta} = X(n)^*DA_{n,\theta}X(n)$ with $D = \partial_i$, ∂^2_{ij}, we only need to study $Y = X(n)^*BX(n)$, under the assumptions "B is symmetric and $b(B) \leqslant R$."

The vector $V = [T_n(2\pi f_{\theta_0})]^{-1/2}X(n)$ is standard gaussian, whence immediately

$$Y = \sum_{k=1}^{n} \tau_k W_k^2$$

where the τ_k are the eigenvalues of

$$C = [T_n(2\pi f_{\theta_0})]^{1/2}B[T_n(2\pi f_{\theta_0})]^{1/2}$$

and where the vector W with coordinates W_k is standard gaussian, and is obtained from V by an isometry of $\mathbb{R}^n$. Consequently

$$E_{P_{\theta_0}}[Y^2] = \sum_{k=1}^{n} 3\tau_k^2 + \sum_{k\neq \ell} \tau_k\tau_\ell = 2\sum_k \tau_k^2 + \left(\sum_k \tau_k\right)^2$$

$$= 2\,\mathrm{tr}(C^2) + [\mathrm{tr}(C)]^2.$$

But for $\alpha(\log f_\theta)$ bounded, (21) shows that $[T_n(2\pi f_\theta)]^{1/2}$ remains bounded in line-norm, and hence in column-norm, by a finite number ρ. By (18), Chapter 10 this proves that $b(C) \leqslant \rho^2 b(B) \leqslant \rho^2 R$; the inequalities $|\mathrm{tr}(C)| \leqslant b(C)$ and $\mathrm{tr}(C^2) \leqslant b(C^2) \leqslant b(C)^2$ imply then

$$E_{P_\theta}(Y^2) \leqslant 3b(C)^2 \leqslant 3\rho^4 R^2$$

which achieves the proof.

Note that if one assumes only that $\alpha(\log f_\theta)$ is finite, instead of 4.2, one can still conclude that

$$E_{P_{\theta_0}} [Y^2_{n,\theta}]$$

remains bounded as (n,θ) varies, which proves (7) as stated above.

4.8. The Asymptotic Information Matrix (Proof of Theorem 4.3)

The density of $X(n)$ being equal to $\exp L_n(\theta,x)$, $x \in \mathbb{R}^n$, one has

$$\int_{\mathbb{R}^n} \exp[L_n(\theta,x)]dx \equiv 1$$

and by derivation under the integral sign

$$E_{P_\theta}[\partial_i L_n(\theta,X)] = \int_{\mathbb{R}^n} (\partial_i L_n)\exp L_n dx \equiv 0.$$

This implies

$$J_n(\theta) = \mathrm{cov}\left[\frac{\partial L_n}{\partial\theta}(\theta,X)\right].$$

Set then

$$\bar{J}_n(\theta) = \mathrm{cov}\left[\frac{\partial \bar{L}_n}{\partial\theta}(\theta,X)\right].$$

But the random part of $(1/n)(\partial\bar{L}_n/\partial\theta)$ may be written

$$\frac{1}{4\pi} I_n\left(\frac{1}{f_\theta^2}\frac{\partial f_\theta}{\partial\theta}\right)$$

and hence

$$\frac{1}{n}\bar{J}_n(\theta) = n \operatorname{cov}\left[\frac{1}{n}\frac{\partial \bar{L}_n}{\partial \theta}\right] = \frac{n}{(4\pi)^2}\operatorname{cov}\left[I_n\left[\frac{1}{f_\theta^2}\frac{\partial f_\theta}{\partial \theta}\right]\right].$$

By Theorem 3.1, Chapter 10 we know that

$$\lim_{n\to\infty} n \operatorname{cov} I_n(\varphi) = 4\pi \int_{\mathbb{T}} \varphi\varphi^* f_\theta^2 \, d\lambda$$

whence the existence of the limit

$$J(\theta) = \lim_{n\to\infty}\frac{1}{n} J_n(\theta) = \lim_{n\to\infty}\frac{1}{n}\bar{J}_n(\theta)$$

$$= \frac{1}{4\pi}\int_{\mathbb{T}} \frac{1}{f_\theta^2}\frac{\partial f_\theta}{\partial \theta}\left[\frac{\partial f_\theta}{\partial \theta}\right]^* d\lambda.$$

Denote now by θ_0, instead of θ, the true value of the parameter of the law of X. For $D = \partial_{ij}^2$ and $\theta \in \Theta$ one has

$$\frac{1}{n} D\bar{L}_n(\theta,X) = -\frac{1}{4\pi}\left[\int_{\mathbb{T}} D(\log f_\theta)d\lambda + I_n\left[D\left[\frac{1}{f_\theta}\right]\right]\right]$$

and hence (Theorem 3.1, Chapter 10), P_{θ_0}-almost surely, the sequence $[-(1/n)D\bar{L}_n(\theta,X)]$ converges, as $n \to \infty$, to $\Phi_{ij}(\theta_0,\theta)$ given by

$$\Phi_{ij}(\theta_0,\theta) = \frac{1}{4\pi}\int_{\mathbb{T}}\left[D(\log f_\theta) + f_{\theta_0} D\left[\frac{1}{f_\theta}\right]\right]d\lambda$$

where the operator D is computed at the point θ.

Call $\Phi(\theta_0,\theta)$ the matrix having generic elements $\Phi_{ij}(\theta_0,\theta)$, which is easily writen in the more compact form

$$\Phi(\theta_0,\theta) = J(\theta) + \frac{1}{4\pi}\int_{\mathbb{T}}\left[1 - \frac{f_{\theta_0}}{f_\theta}\right]\left[\frac{f''_\theta}{f_\theta} - 2\frac{f'_\theta f'^*_\theta}{f_\theta^2}\right]d\lambda. \tag{23}$$

In particular $\Phi(\theta_0,\theta)$ is continuous in θ and $\Phi(\theta_0,\theta_0) \equiv J(\theta_0)$. Thus we have P_{θ_0}-a.s.,

$$\lim_{n\to\infty}\left[-\frac{1}{n}\frac{\partial^2\bar{L}_n}{\partial\theta^2}(\theta,X)\right] = \Phi(\theta_0,\theta).$$

By 4.7,

$$\sup_{\theta\in\Theta}\left\|\frac{\partial^2}{\partial\theta^2}(L_n - \bar{L}_n)\right\|$$

remains bounded in L^2-norm (under the law P_{θ_0}) as n varies. By Borel-Cantelli's lemma we conclude that there is a *finite* random integer $N(\omega)$ such that P_{θ_0}-a.s. one has

$$\left\| -\frac{1}{n}\frac{\partial^2 \overline{L}_n}{\partial\theta^2} + \frac{1}{n}\frac{\partial^2 L_n}{\partial\theta^2} \right\| \leqslant \frac{1}{n^{1/4}} \qquad \text{for } n \geqslant N(\omega).$$

Hence P_{θ_0}-a.s., the sequence

$$\left[-\frac{1}{n}\frac{\partial^2 L_n}{\partial\theta^2}(\theta,X) \right]$$

converges as well to $\Phi(\theta_0,\theta)$. Same result for $\widetilde{L}_n$ by 4.7. This concludes the proof of Theorem 4.3.

4.9. Remark. As in 2.3, one proves separately that

$$\left[-\frac{1}{n}\frac{\partial^2 \overline{L}_n}{\partial\theta^2}(\theta,X) \right] \quad \text{and} \quad \left[-\frac{1}{n}\frac{\partial^2 L_n}{\partial\theta^2}(\theta,X) \right]$$

are P_{θ_0}-a.s. equicontinuous functions of $\theta \in \Theta$. *This implies*

(same argument as in 2.3) that P_{θ_0}*-a.s., the convergence of*

$$\left[-\frac{1}{n}\frac{\partial^2 \overline{L}_n}{\partial\theta^2}(\theta,X) \right] \quad \text{and} \quad \left[-\frac{1}{n}\frac{\partial L_n}{\partial\theta^2}(\theta,X) \right]$$

to $\Phi(\theta_0,\theta)$ *is uniform in* $\theta \in \Theta$. *The same result is true for* $\widetilde{L}_n$ *by* 4.7.

4.10. Asymptotic Normality (Proof of Theorem 4.4)

Assume θ to be *interior* and *injective* in Θ. Let θ_n be an estimator associated to L_n; we know that P_θ-a.s., θ_n converges towards θ as $n \to \infty$. For n large, θ_n is hence interior to Θ and hence $(\partial/\partial\theta)L_n(\theta_n,X)$ is zero. Taylor's formula grants then the existence of $u \in [0,1]$ such that $\theta'_n = u\theta + (1 - u)\theta_n$ satisfies

$$(24) \qquad \frac{\partial L_n}{\partial\theta}(\theta,X) + \frac{\partial^2 L_n}{\partial\theta^2}(\theta'_n,X)\cdot[\theta_n - \theta] = 0.$$

Assume that $J(\theta)$ is *invertible*. Then P_θ-a.s.,

$$\left[-\frac{1}{n}\frac{\partial^2 L_n}{\partial\theta^2}(\theta',X) \right]$$

converges (uniformly in θ') to $\Phi(\theta,\theta')$ and θ'_n converges to θ, whence the P_θ-a.s. convergence of

$$M_n = -\frac{1}{n}\frac{\partial^2 L_n}{\partial\theta^2}(\theta'_n,X)$$

to $\Phi(\theta,\theta) = J(\theta)$. In particular for n large, M_n is invertible and by (24) one has

(25) $$\sqrt{n}(\theta_n - \theta) = M_n^{-1} \frac{1}{\sqrt{n}} \frac{\partial L_n}{\partial \theta} (\theta,X).$$

It is now sufficient to prove that the law of

$$\frac{1}{\sqrt{n}} \frac{\partial L_n}{\partial \theta}(\theta,X)$$

converges to $N(0,J(\theta))$ to conclude that the law of $\sqrt{n}(\theta_n - \theta)$ converges to $N[0,J(\theta)^{-1}]$.

The same argument will then apply to $(\bar{L}_n,\bar{\theta}_n)$ and $(\breve{L}_n,\breve{\theta}_n)$, by 4.8, 4.9 for the convergence of $\bar{M}_n$, $\breve{M}_n$ to $J(\theta)$, and by 4.7, 4.6 to make sure that the laws of

$$\frac{1}{\sqrt{n}} \frac{\partial \bar{L}_n}{\partial \theta}, \quad \frac{1}{\sqrt{n}} \frac{\partial L_n}{\partial \theta}, \quad \frac{1}{\sqrt{n}} \frac{\partial \breve{L}_n}{\partial \theta}$$

have the same limit. In view of this last point, we only need to study

$$\bar{Y}_n = \frac{1}{\sqrt{n}} \frac{\partial \bar{L}_n}{\partial \theta}(\theta,X)$$

which by (5) may be written

$$\bar{Y}_n = \frac{\sqrt{n}}{4\pi} \left[-\int_{\mathbb{T}} \frac{1}{f_\theta} \frac{\partial f_\theta}{\partial \theta} d\lambda + I_n\left[\frac{1}{f_\theta^2} \frac{\partial f_\theta}{\partial \theta}\right]\right]$$

(26)

$$= \sqrt{n}[I_n(\varphi) - I(\varphi)]$$

with $$\varphi = \frac{1}{4\pi f_\theta^2} \frac{\partial f_\theta}{\partial \theta} .$$

By Theorem 3.1, Chapter 10 we see that the distribution of Y_n converges to $N(0,\Gamma)$ where

$$\Gamma = 4\pi \int_{\mathbb{T}} \varphi\varphi^* f_\theta^2 \, d\lambda$$

is clearly equal to $J(\theta)$, which achieves the proof of Theorem 4.4.

Bibliographical Hints

The questions studied in this chapter have received considerable attention in the literature. The precise study of Whittle's approximation within the 0(1)-approximation is due to *Phan-Dinh*, *Coursol* and *Dacunha-Castelle*. The results on asymptotic efficiency can be found in *Hannan*, under different hypotheses.

Chapter XIV
IDENTIFICATION AND COMPENSATED LIKELIHOOD

1. Identification

Let X be a centered, stationary gaussian process, with unknown spectral density. We are going to assume that X can be considered as an ARMA process with unknown order (p,q). Heuristically this methodological choice relies on two remarks: the rational densities are dense, in various senses, in the set of all densities, and the interpretation of ARMA parameters is fairly intuitive; moreover the actual modelization within the ARMA family can be completed with a reasonable amount of computation, and often with a rather small number of parameters.

To go from the observations to a specific ARMA model, the first step is the use of various empirical estimation techniques to select a type of model (ARMA(p,q), or ARIMA, or a model with seasonal effects). These methods do not lead to a clearcut choice for the type of model, or its dimension (p,q). This first step can be called **empirical identification**, and deals with the choice of model type and dimensional parameters. The second step is the precise estimation of the parameters within a class of ARMA models of fixed type and fixed dimension.

After having used likelihood techniques for the estimation stage, we are going to use them for the identification itself.

2. Parametrization

2.1. Parameter Spaces and Notations

Let X be a centered gaussian ARMA process with innovation W. We shall parametrize the spectral density

$$f_\theta(\lambda) = \frac{\sigma^2}{2\pi}\left|\frac{Q}{P}(e^{-i\lambda})\right|^2$$

of X by $\theta = (P,Q,\sigma)$ as in Chapter 13, Section 3.1. More precisely, *call* $\mathcal{P}_r$ *the r-dimensional affine space of all polynomials* R, *with real coefficients, such that* $R(0) = 1$ *and* $d^\circ R \leq r$. Fix the dimensional parameter $\nu = (p,q)$. The equivalence classes of ARMA(p,q) invertible processes correspond exactly to the set $B(\nu) \subset \mathcal{P}_p \times \mathcal{P}_q \times \mathbb{R}^+$ of all $\theta = (P,Q,\sigma)$ such that PQ has all its roots of modulus > 1, Q/P is irreducible, and $d^\circ P = p$, $d^\circ Q = q$.

However $B(\nu)$ is not compact and in standard estimation practice the parameter set is, *implicitly at least*, a compact subset of $B(\nu)$. Moreover ν is not going to be considered as known here, so we shall have to allow for simultaneous consideration of polynomials P, Q with degrees $\leq p,q$ and the possible use of reducible fractions Q/P. This leads to the definition of a whole family $\Theta(\nu,\rho,u)$ of compact parameter spaces where $\nu = (p,q)$, $\rho > 0$, $u > 1$.

Define first *the set* $\mathcal{P}_r(\rho)$ *of all* $R \in \mathcal{P}_r$ *having all their roots of modulus* $\geq 1 + \rho$. Clearly, this is a compact subset of $\mathcal{P}_r$. We then let

$$\Theta(\nu,\rho,u) = \mathcal{P}_p(\rho) \times \mathcal{P}_q(\rho) \times \left[\frac{1}{u}, u\right].$$

Note that *several points* $\theta \in \Theta(\nu,\rho,u)$ *may represent the same equivalence class of ARMA processes*. In fact two points θ, $\tilde{\theta}$ in $\Theta(\nu,\rho,u)$ have this property if and only if $f_\theta \equiv f_{\tilde{\theta}}$, that is if and only if $\sigma = \tilde{\sigma}$ and $Q/P \equiv \tilde{Q}/\tilde{P}$.

For $\nu = (p,q)$, $\nu' = (p',q')$ we shall say that $\nu \leq \nu'$ if $\{p \leq p'$ and $q \leq q'\}$. Since $\mathcal{P}_p(\rho) \subset \mathcal{P}_{p'}(\rho)$ one has then clearly $\Theta(\nu,\rho,u) \subset \Theta(\nu',\rho,u)$ for fixed ρ,u and $\nu \leq \nu'$.

Consider the standard situation where the actual order ν_0 of an ARMA process X is unknown. One has to estimate ν_0 and $\theta \in \Theta(\nu_0,\rho,u)$. In practice one may always assume that $\nu_0 \leq \nu_M$ *where* ν_M *as well as* ρ *and* u *are imposed by concrete computing limitations*, and hence can be considered as **known**.

From now on we shall hence consider ρ, u and ν_M as given. We shall then write $\Theta(\nu)$ instead of $\Theta(\nu,\rho,u)$.

2.2. Necessity of a Parsimony Principle

Call $\bar{L}_n(\theta)$ the log-likelihood of the first n observations

$$X(n) = \begin{bmatrix} X_1 \\ \cdot \\ \cdot \\ \cdot \\ X_n \end{bmatrix}$$

under the assumption that X is gaussian and has spectral density f_θ. Note that $L_n(\theta)$ depends only on f_θ. Call P_θ the law of X.

Define random variables $\theta_n(\nu) \in \Theta(\nu)$ and $L_{n,\nu} \in \mathbb{R}$ by

$$L_{n,\nu} = \sup_{\theta \in \Theta(\nu)} L_n(\theta) = L_n[\theta_n(\nu)].$$

One could attempt to estimate the true dimension ν_0 by a random variable ν_n such that $0 \leq \nu_n \leq \nu_M$ and

$$L_{n,\nu_n} = \sup_{0 \leq \nu \leq \nu_M} L_{n,\nu}.$$

However for large n the estimator ν_n is always equal to ν_M *and hence is not consistent.* If one does not impose any bound ν_M, the situation is worse since the estimator ν_n then tends to $+\infty$ as $n \to +\infty$.

In fact, overall maximization of the likelihood in (ν,θ) leads to a model whose absolute fit with the observations may *seem* remarkably good, *but for ν large the simultaneous errors on the large number of parameters destroy seriously the predictive quality of the model.*

Hence quality of fit and quality of prediction cannot both be improved indefinitely. One must seek a compromise between the two, which will rely on a heuristic **parsimony principle** favoring small dimensional models. To apply this principle we shall add to the usual log-likelihood *a* **compensator** which will decrease the "meaningful" likelihood when the dimension increases. *The goal being to obtain consistent estimators of ν_0*, thus realizing the compromise just mentioned.

3. Compensated Likelihood

3.1. Definitions. We call **compensator** any deterministic sequence of functions δ_n: $\mathbb{N} \times \mathbb{N} \to \mathbb{R}^+$ such that $\nu \leqslant \nu'$ implies $\delta_n(\nu) \leqslant \delta_n(\nu')$. **The compensated log-likelihood** is defined by

$$A_n(\nu,\theta) = L_n(\theta) - \delta_n(\nu) \quad \text{for } \theta \in \Theta(\nu).$$

A maximum compensated likelihood estimator will be any pair of estimators $\hat{\nu}_n$, $\hat{\theta}_n$ such that $\hat{\theta}_n \in \Theta(\hat{\nu}_n)$ and

$$A_n(\hat{\nu}_n,\hat{\theta}_n) = \sup_{0 \leqslant \nu \leqslant \nu_M} \sup_{\theta \in \Theta(\nu)} A_n(\nu,\theta).$$

It will be called consistent if $\hat{\nu}_n \to \nu_0$ and $\hat{\theta}_n \to \theta_0$, P_{θ_0}-a.s. as $n \to \infty$ where ν_0, θ_0 is the true model. Note that ν_M being finite, this means that P_{θ_0}-a.s. $\hat{\nu}_n$ is equal to ν_0 for n large enough.

3.2. The Choice of a Compensator

There are two types of errors of identification, *overparametrization* and *underparametrization* which occur respectively when $\hat{\nu}_n > \nu_0$ or $\hat{\nu}_n < \nu_0$.

(a) Let $\theta_n(\nu)$ be an ordinary maximum likelihood estimator in $\Theta(\nu)$. We shall see below that for $\nu < \nu_0$,

$$L_n[\theta_n(\nu)] - L_n[\theta_n(\nu_0)] \sim -n\gamma$$

with $\gamma > 0$. To avoid the selection of such a ν, the difference of compensators $[-\delta_n(\nu) + \delta_n(\nu_0)]$, which is positive, should be $<< n\gamma$. But γ is in fact unknown and arbitrary. Hence it is practically necessary to impose the condition

$$\lim_{n\to\infty} \frac{\delta_n(\nu)}{n} = 0$$

to avoid underparametrization.

(b) Assume that $\nu_0 \leqslant \nu$. One can prove in this case that for n large

$$E[L_n[\theta_n(\nu)] - L_n(\theta_0)] \simeq r$$

where $r = p + q - (p - p_0) \wedge (q - q_0)$ and $\nu_0 = (p_0,q_0)$, $\nu = (p,q)$.

This has suggested the use of the compensators $\delta_n(\nu) = c|\nu|$ where $|\nu| = p + q$ for $\nu = (p,q)$. Here c is a "suitable" constant. Such a choice, strongly recommended by Akaike in his pioneering articles, has later been recognized as *unsatisfactory since it leads in fact to a strictly positive probability of overparametrization.*

We shall see below that the choices $\delta_n(\nu) = (1/2)|\nu|(1 + |\nu|)\varphi(n)$ with

$$\lim_{n\to\infty} \frac{\varphi(n)}{n} = 0 \quad \text{and} \quad \underline{\lim}_{n\to\infty} \frac{\varphi(n)}{\log\log n} > 1$$

are sufficient (and "practically" necessary) to grant the consistency of $\hat{\nu}_n$. The arguments below *seem to recommend* the choice of the smallest possible consistent compensator, that is $\varphi(n) = (1 + \varepsilon)\log\log n$ with small $\varepsilon > 0$. This will however not be proved rigorously. In fact other considerations have led to other choices such as $\delta_n(\nu) = |\nu|\log n$ (cf. [44]).

4. Mathematical Study of Compensated Likelihood

4.1. Hypothesis

Given $\rho > 0$, $u > 1$ and a maximal order ν_M we set $\Theta(\nu) = \Theta(\nu,\rho,u)$ with $0 \leqslant \nu \leqslant \nu_M$. We assume that X is a gaussian centered ARMA process of exact order $\nu_0 \leqslant \nu_M$. The true spectral density of X is denoted by f_{θ_0} with $\theta_0 = (P_0, Q_0, \sigma_0)$ in $\Theta(\nu_0)$ and $\nu_0 = (p_0,q_0)$, $d^\circ P_0 = p_0$, $d^\circ Q_0 = q_0$ and Q_0/P_0 **irreducible**. In particular θ_0 is necessarily an **injective** point of $\Theta(\nu_0)$ (cf. Chapter 13, Section 3.1). Moreover we assume θ_0 to be an **interior** point of $\Theta(\nu_0)$, which simply means here that $1/u < \sigma_0 < u$ and that P_0Q_0 has all its roots of modulus $> 1+\rho$.

We shall often find convenient to use the following notation, for $\theta \in \Theta(\nu)$ and $\nu = (p,q)$,

$$\theta = (\beta,\sigma) \quad \text{with} \quad \beta = (P,Q) \in \mathcal{P}_p \times \mathcal{P}_q.$$

4.2. Maximum Compensated Likelihood Estimators

Consider the log-likelihood $L_n(\theta)$ of X and a given compensator $\delta_n(\nu)$. Define estimators $\theta_n(\nu) \in \Theta(\nu)$, $\nu_n \in [0,\nu_M]$, and the maximal log-likelihood $L_{n,\nu}$ by

(1) $L_{n,\nu} = \sup_{\theta \in \Theta(\nu)} L_n(\theta) = L_n[\theta_n(\nu)]$

(2) $A_n(\nu) = L_{n,\nu} - \delta_n(\nu)$

(3) $A_n(\nu_n) = \sup_{0 \leqslant \nu \leqslant \nu_M} A_n(\nu).$

Of course if one replaces the exact log-likelihood $\overline{L}_n$ by its standard approximations $\underline{L}_n$, $\tilde{L}_n$ (cf. Chapter 13, Section 1.1), one can construct similarly estimators $\overline{\theta}_n(\nu)$, $\overline{\nu}_n$ and $\hat{\theta}_n(\nu)$, $\hat{\nu}_n$ having the same asymptotic behavior as $\theta_n(\nu)$, ν_n as will be seen below.

4.3. Theorem. *Assume hypothesis* 4.1 *and write* $|\nu| = p + q$ *for* $\nu = (p,q)$. *Select any compensator of the form*

$$\delta_n(\nu) = \frac{1}{2}|\nu|(1 + |\nu|)\varphi(n)$$

where the function φ: $\mathbb{N} \to \mathbb{R}^+$ *satisfies*

$$\lim_{n\to\infty} \frac{\varphi(n)}{n} = 0 \qquad \underline{\lim}_{n\to\infty} \frac{\varphi(n)}{\log \log n} > 1.$$

Then the corresponding maximum compensated likelihood estimator ν_n *converges* P_{θ_0}-a.s. *to the true order* ν_0. *Same result when the exact likelihood is replaced by its standard approximations* $\underline{L}_n$, $\tilde{L}_n$.

Proof. This is a direct consequence of the more technical results 4.5, 4.6, and 4.7. The deepest of these results uses a rather delicate estimate 4.8 for the maximal log-likelihood $(L_{n,\nu} - L_{n,\nu_0})$, which in turn relies on sophisticated versions of the law of the iterated logarithm for the periodogram.

4.4. Remark. When φ is too large, that is when

$$\underline{\lim}_{n\to\infty} \frac{\varphi(n)}{n} > 0,$$

consistency of ν_n **fails** with probability 1, provided ν_0 happens to be large enough.

When φ is too small, that is when

$$\overline{\lim_{n\to\infty}} \frac{\varphi(n)}{\log \log n} < 1,$$

one can actually supply examples where consistency fails with strictly positive probability.

Hence Theorem 4.3 is sharp, and in an obvious sense (cf. 4.6, 4.7) exhibits *the smallest consistent compensators.*

4.5. Theorem (Compensators Avoiding Underparametrization). *Assume hypothesis* 4.1. *Let* δ_n *be a compensator such that*

$$\lim_{n\to\infty} \frac{\delta_n(\nu)}{n} = 0$$

for all $\nu \in \mathbb{N} \times \mathbb{N}$. *Then by overall maximization of the compensated log-likelihood, one avoids asymptotic underparametrization* P_{θ_0}-a.s. *This means that*

$$\underline{\lim}_{n\to\infty} \nu_n \geqslant \nu_0 \qquad P_{\theta_0}\text{-a.s.}$$

Same result when the exact log-likelihood L_n *is replaced by* $\bar{L}_n$, $\tilde{L}_n$.

Proof. Fix $\nu < \nu_0$. One has then $\theta_0 \notin \Theta(\nu)$, and hence the Kullback information K satisfies

$$0 < \gamma = \inf_{\theta\in\Theta(\nu)} K(P_{\theta_0}, P_\theta).$$

The hypotheses of Chapter 13, Theorem 2.3 obviously hold for the densities f_θ, $\theta \in \Theta(\nu)$. Hence P_{θ_0}-a.s., for every fixed ν, the sequence of functions

$$\psi_n(\theta) = \frac{1}{n}[L_n(\theta_0) - L_n(\theta)]$$

converge *uniformly in* $\theta \in \Theta(n)$ to the limit $\psi(\theta) = K(P_{\theta_0}, P_\theta)$. But this always implies

$$\lim_{n\to\infty} [\inf_{\theta\in\Theta(\nu)} \psi_n(\theta)] = \inf_{\theta\in\Theta(\nu)} \psi(\theta)$$

so that P_{θ_0}-a.s.

$$(4) \qquad \gamma = \lim_{n\to\infty} \frac{1}{n} [L_n(\theta_0) - \sup_{\theta\in\Theta(\nu)} L_n(\theta)].$$

Since θ_0 is an injective point of $\Theta(\nu_0)$, Theorem 3.2, Chapter 13 implies

$$\lim_{n\to\infty} \theta_n(\nu) = \theta_0, \quad P_{\theta_0}\text{-a.s.}$$

The P_{θ_0}-a.s. uniform convergence of the ψ_n forces then, P_{θ_0}-a.s.

(5) $$\lim_{n\to\infty} \frac{1}{n}[L_n(\theta_0) - L_n(\theta_n(\nu_0))] = \lim_{n\to\infty} \psi_n[\theta_n(\nu_0)] = \psi(\theta_0) = 0.$$

By definition (1) we have

$$L_{n,\nu} = \sup_{\theta\in\Theta(\nu)} L_n(\theta) \quad \text{and} \quad L_{n,\nu_0} = L_n[\theta_n(\nu_0)]$$

and hence (4) and (5) imply P_{θ_0}-a.s.

$$\lim_{n\to\infty} \frac{1}{n} [L_{n,\nu} - L_{n,\nu_0}] = -\gamma \qquad \text{for } \nu < \nu_0$$

which, by definition (2) of $A_n(\nu)$ and the hypothesis on δ_n, implies P_{θ_0}-a.s.,

$$\lim_{n\to\infty}[A_n(\nu) - A_n(\nu_0)] = -\infty \qquad \text{for } \nu < \nu_0.$$

Since, by definition (3), $A_n(\nu_n) - A_n(\nu_0) \geqslant 0$ we conclude that P_{θ_0}-a.s all the limiting values of ν_n must belong to $[\nu_0,\nu_M]$.

An identical argument handles the case of $\bar{L}_n$, $\hat{L}_n$.

4.6. Theorem (Compensators Avoiding Overparametrization). *Assume hypothesis* 4.1. *Consider two functions* h: $\mathbb{N} \times \mathbb{N} \to \mathbb{R}^+$ *and* φ: $\mathbb{N} \to \mathbb{R}^+$ *such that*

(6) $$\underline{\lim}_{n\to\infty} \frac{\varphi(n)}{\log\log n} > 1$$

(7) $$h(p',q') - h(p,q) \geqslant p' + q' - [(p'-p) \wedge (q'-q)]$$

$$\text{for } 0 \leqslant p \leqslant p',\ 0 \leqslant q \leqslant q'.$$

Consider the compensator $\delta_n(\nu) = h(\nu)\varphi(n)$. *Then maximization of the compensated log-likelihood provides estimators* ν_n *of* ν *which* P_{θ_0}-a.s. *avoid overparametrization. This means that*

$$\overline{\lim_{n\to\infty}}\, \nu_n \leqslant \nu_0 \qquad P_{\theta_0}\text{-a.s.}$$

Same result when the exact likelihood L_n *is replaced by* $\bar{L}_n$, $\hat{L}_n$.

Proof. Assume that $\nu_0 < \nu$. The assumed expression of δ_n and definition (2) for $A_n(\nu)$ give

$$A_n(\nu) - A_n(\nu_0) = L_{n,\nu} - L_{n,\nu_0} - \varphi(n)[h(\nu) - h(\nu_0)].$$

Let $c(\nu,\nu_0) = p + q - [(p - p_0) \wedge (q - q_0)]$ for $\nu = (p,q)$, $\nu_0 = (p_0,q_0)$. By hypothesis, $h(\nu) - h(\nu_0) \geqslant c(\nu,\nu_0) > 0$ while the crucial theorem 4.8 below tells us that

$$\overline{\lim_{n\to\infty}} \frac{1}{\log\log n}(L_{n,\nu} - L_{n,\nu_0}) \leqslant c(\nu,\nu_0) \quad P_{\theta_0}\text{-a.s.}$$

We thus obtain P_{θ_0} a.s., for $\nu_0 < \nu$

$$\overline{\lim_{n\to\infty}} \frac{1}{\log\log n}[A_n(\nu) - A_n(\nu_0)]$$

$$\leqslant c(\nu,\nu_0)\left[1 - \underline{\lim_{n\to\infty}} \frac{\varphi(n)}{\log\log n}\right].$$

Hypothesis (6) on φ forces the right-hand side to be strictly negative, and hence

$$\overline{\lim_{n\to\infty}} [A_n(\nu) - A_n(\nu_0)] = -\infty \qquad \text{for } \nu > \nu_0.$$

Since $A_n(\hat{\nu}_n) - A_n(\nu_0) \geqslant 0$ the theorem is proved. An identical proof handles the case of $\overline{L}_n$, $\tilde{L}_n$.

We now elaborate briefly on hypothesis (7).

4.7. Lemma. *If a function* h: $\mathbb{N} \times \mathbb{N} \to \mathbb{R}^+$ *satisfies* (7), *then one necessarily has*

$$h(p,q) \geqslant h(0,0) + \frac{1}{2}(p+q)(p+q+1) \quad \textit{for all } p \geqslant 0,\ q \geqslant 0.$$

On the other hand for any fixed $a \in \mathbb{R}^+$, *the function*

$$(p,q) \to a + \frac{1}{2}(p + q)(p + q + 1)$$

satisfies (7).

Proof. The direct part is proved by double induction on p,q. The converse is elementary. Hence the compensators

$$\frac{1}{2}(p + q)(p + q + 1)\varphi(n)$$

with

$$\underline{\lim}_{n\to\infty} \frac{\varphi(n)}{\log \log n} > 1$$

are the smallest avoiding overparametrization. The key point in the proof of Theorem 4.6 was the following result.

4.8. Theorem. *Assume hypothesis* 4.1. *Let* $L_{n,\nu}$ *be the maximum of the log-likelihood* $L_n(\theta)$ *for* $\theta \in \Theta(\nu)$. *Then for* $\nu = (p,q)$, $\nu_0 = (p_0,q_0)$ *and* $\nu \geqslant \nu_0$ *one has*

$$(8)\quad \overline{\lim}_{n\to\infty} \frac{1}{\log \log n}[L_{n,\nu} - L_{n,\nu_0}] \leqslant p + q - [(p-p_0)\wedge(q-q_0)],$$

P_{θ_0}-a.s. *The same result holds when the exact log-likelihood* L_n *is replaced by its standard approximations* $\bar{L}_n$, $\hat{L}_n$.

Proof. The proof is quite delicate and will only be completed in Section 6.10 below. Many of the technical difficulties stem from the lack of injectivity of the map $\theta \to f_\theta$ on $\Theta(\nu)$.

4.9. Conjecture

We are convinced (and have checked in particular cases) that the $\overline{\lim}_{n\to\infty}$ in Theorem 4.8 is actually equal to $p + q - (p - p_0) \wedge (q - q_0)$. This point is needed to show the nonconsistency of $\delta_n(\nu) = (1/2)(p+q)(p+q+1)\varphi(n)$ as soon as

$$\overline{\lim}_{n\to\infty} \frac{\varphi(n)}{\log \log n} < 1.$$

5. Noninjective Parametrization

5.1. The Main Difficulties

Assume hypothesis 4.1, and fix $\nu = (p,q) > \nu_0 = (p_0,q_0)$. Recall that $\theta_0 = (P_0,Q_0,\sigma_0) = (\beta_0,\sigma_0)$ with Q_0/P_0 irreducible etc. as in 4.1. Let $m = (p - p_0) \wedge (q - q_0)$ and call $M \subset P_p(\rho) \times P_q(\rho)$ the set of all pairs $\beta = (SP_0,SQ_0)$ where the polynomial S is arbitrary in $P_m(\rho)$. The notations are those of Section 2.1.

Then, as seen earlier,

$$(9)\qquad \mathcal{M} \times \{\sigma_0\} = \{\theta \in \Theta(\nu) \mid f_\theta = f_{\theta_0}\} = \mathcal{N}(\theta_0).$$

It is intuitively clear that S cannot be estimated consistently since the whole history of X can only determine f_{θ_0}. Thus $S \in \mathcal{P}_m(\rho)$ plays the part of a *nuisance parameter.* The ordinary maximum likelihood estimator $\theta_n(\nu) = (\beta_n(\nu),\sigma_n(\nu))$ in $\Theta(\nu)$ is generally not consistent. Here $\beta_n(\nu) \in \mathcal{P}_p(\rho) \times \mathcal{P}_q(\rho)$ and $\sigma_n(\nu) \in \mathbb{R}^+$. The only consistency properties which hold P_{θ_0}-a.s. by Chapter 13, Section 3.2 are the following:

$$(10)\qquad \lim_{n\to\infty} f_{\theta_n(\nu)} = f_{\theta_0} \qquad P_{\theta_0}\text{-a.s.}$$

where convergence is understood in the sense of $\| \ \|_\infty$ on $[-\pi,\pi]$, and

(11) the distance between $\theta_n(\nu)$ and $\mathcal{N}(\theta_0)$ tends to zero P_{θ_0}-a.s.

This is obviously equivalent to

(12) the distance between $\beta_n(\nu)$ and $\mathcal{M}$ tends to zero P_{θ_0}-a.s.

$$(13)\qquad \lim_{n\to\infty} \sigma_n(\nu) = \sigma_0, \quad P_{\theta_0}\text{-a.s.}$$

Another source of trouble is that by Chapter 12, Section 3.3, as soon as $p > p_0$ and $q > q_0$, the information matrix $J(\theta)$ is *not invertible* for $\theta \in \mathcal{N}(\theta_0)$. To deal with the two problems just stated we introduce a new system of coordinates in $\Theta(\nu)$, linked to the point θ_0.

5.2. A New System of Coordinates

Notations and assumptions are those of the preceding paragraph.

Let $\mathcal{U}$ be the _vector_ space of all pairs $U = (\overline{P},\overline{Q})$ of polynomials with $\overline{P}(0) = \overline{Q}(0) = 0$, such that

$$(14)\qquad d^\circ\overline{P} \leq p_0,\ d^\circ\overline{Q} \leq q \qquad \text{if } p - p_0 \leq q - q_0$$

$$(15)\qquad d^\circ\overline{P} \leq p,\ d^\circ\overline{Q} \leq q_0 \qquad \text{if } p - p_0 > q - q_0.$$

From now on, *in all proofs we shall always assume implicitly* that (14) holds, the case of (15) being handled completely symetrically.

Let $m = (p - p_0) \wedge (q - q_0)$. There is then *an affine bijection* G: $\mathcal{P}_m \times \mathcal{U} \to \mathcal{P}_p \times \mathcal{P}_q$ *defined as follows*: for $S \in \mathcal{P}_m$, $U = (\bar{P},\bar{Q}) \in \mathcal{U}$, we set $G(S,U) = (P,Q)$ where

$$(16) \qquad P = P_0 S + \bar{P}, \qquad Q = Q_0 S + \bar{Q}$$

We check first that G is one-to-one; if $U = (\bar{P},\bar{Q})$ and $U' = (\bar{P}',\bar{Q}')$, the relation $G(S,U) = G(S',U')$ implies $\bar{P}' - \bar{P} = P_0(S-S')$; by (14) $\bar{P}$, $\bar{P}'$, and P_0 have the same degree. Then $S - S'$ is a constant polynomial, which must be 0 since $S(0) = S'(0) = 1$. This gives immediately $S = S'$, $U = U'$.

To see that G is onto, start with $(P,Q) \in \mathcal{P}_p \times \mathcal{P}_q$. By euclidean division we write $P = \Delta P_0 + R$ with $d°R < p_0$. Set now

$$\bar{P} = R + (\Delta(0) - 1)P_0, \quad S = \Delta - \Delta(0) + 1, \quad \bar{Q} = Q - Q_0 S$$

to check immediately that $G(S,U) = (P,Q)$ and $(S,U) \in \mathcal{P}_m \times \mathcal{U}$ provided $p - p_0 \leqslant q - q_0$. Symmetric argument when $p - p_0 > q - q_0$.

Our new coordinate system (S,U) *depends of course on* θ_0, but will only be used for theoretical arguments, and certainly not for actual estimation. Actually these coordinates will essentially be used in a small neighbourhood of the "fiber" $\mathcal{M}$ at θ_0 introduced in (9).

5.3. A Small Neighborhood of $\mathcal{M}$

By definitions (9) and (16), the "fiber" $\mathcal{M}$ at θ_0 satisfies

$$(17) \qquad \mathcal{M} = G(\mathcal{P}_m(\rho) \times \{0\}) \subset \mathcal{P}_p(\rho) \times \mathcal{P}_q(\rho).$$

Moreover $\mathcal{P}_j(\rho')$ is by construction a compact neighborhood of $\mathcal{P}_j(\rho)$ whenever $0 < \rho' < \rho$. Since G is a homeomorphism there is a small convex compact neighborhood $\mathcal{W}_0$ of 0 in $\mathcal{U}$ such that for any other convex compact neighborhood $\mathcal{W}$ of 0 in $\mathcal{U}$, the inclusion $\mathcal{W} \subset \mathcal{W}_0$ implies

$$(18) \qquad G\left[\mathcal{P}_m\left(\frac{\rho}{2}\right) \times \mathcal{W}\right] \subset \mathcal{P}_p\left(\frac{\rho}{3}\right) \times \mathcal{P}_q\left(\frac{\rho}{3}\right).$$

We shall write

(19) $\quad S = \mathcal{P}_m\left[\frac{\rho}{2}\right], \quad K = G(S \times \mathcal{W}), \; V = K \cap [\; \mathcal{P}_p(\rho) \times \mathcal{P}_q(\rho)].$

Note that V is a compact neighborhood of $\mathcal{M}$ in $\mathcal{P}_p(\rho) \times \mathcal{P}_q(\rho)$.

The consistency property (12) now implies that the maximum likelihood estimator $\theta_n(\nu) = (\beta_n(\nu), \sigma_n(\nu))$ satisfies

(20) $\quad \beta_n(\nu) \in V \subset K = G(S \times \mathcal{W}) \qquad \text{for } n \geqslant N(\omega)$

where $N(\omega)$ is a P_{θ_0}-a.s. finite random integer.

6. Almost Sure Bounds for the Maximal Log-Likelihood

6.1. Notations

We now undertake the proof of Theorem 4.8. Hypotheses and notations are those of Sections 2.1, 4.1 and 5.1. However we shall lighten these notations as follows: in the whole paragraph 6 *the dimensional parameter* $\nu = (p,q) \geqslant \nu_0$ *is fixed and will most of the time be omitted in the notations.* We shall also set $\mathcal{B} = \mathcal{P}_p(\rho) \times \mathcal{P}_q(\rho)$, so that $\Theta(\nu) \subset \mathcal{B} \times \mathbb{R}^+$, and we systematically write $\theta = (\beta,\sigma)$ with $\beta = (P,Q) \in \mathcal{B}$ whenever $\theta \in \Theta(\nu)$.

Call L_n, $\breve{L}_n$, $\bar{L}_n$ the log-likelihood and its standard approximations (Chapter 13, Section 1.1). *Let ℓ_n be anyone of the three functions L_n, $\breve{L}_n$, $\bar{L}_n$.* To ℓ_n we associate an estimator $\theta_n = (\beta_n, \sigma_n) \in \Theta(\nu)$ such that

(21) $\quad \ell_{n,\nu} = \sup_{\theta \in \Theta(\nu)} \ell_n(\theta) = \ell_n(\theta_n).$

We want to study

$$\overline{\lim_{n\to\infty}} \frac{1}{\log \log n}(\ell_{n,\nu} - \ell_{n,\nu_0}).$$

6.2. Reduction to the Case of Approximate Log-Likelihood

From Chapter 13, Section 4.6 we get a constant c such that

$$\sup_n \sup_{\theta \in \Theta(\nu)} |L_n(\theta) - \breve{L}_n(\theta)| \leqslant c$$

so that to study

$$\frac{1}{\log\log n}(\ell_{n,\nu} - \ell_{n,\nu_0})$$

it will obviously be sufficient to consider only the cases $\ell_n = \tilde{L}_n$ *and* $\ell_n = \bar{L}_n$, a restriction which will be assumed from now on.

6.3. Reduction to the Case of Constant Variance

For $\beta = (P,Q) \in \mathcal{B}$ let

$$(22) \qquad g_\beta(\lambda) = \left|\frac{Q}{P}(e^{-i\lambda})\right|^2 \qquad -\pi \leqslant \lambda \leqslant \pi.$$

Then by formula (18) of Chapter 12, the integral $\int_{\Pi} \log g_\beta d\lambda$ is zero and formulas (3), (4) and (5), Chapter 13 then become

$$(23) \qquad \tilde{L}_n(\beta,\sigma) = -\frac{1}{2}\left[n\log(2\pi\sigma^2) + \frac{1}{\sigma^2}X(n)^*[T_n(g_\beta)]^{-1}X(n)\right]$$

$$(24) \qquad \bar{L}_n(\beta,\sigma) = -\frac{1}{2}\left[n\log(2\pi\sigma^2) + \frac{1}{\sigma^2}X(n)^*T_n\left(\frac{1}{g_\beta}\right)X(n)\right]$$

where $T_n(\cdot)$ is the Toeplitz matrix. Thus both $\tilde{L}_n$, $\bar{L}_n$ are of the form

$$\ell_n(\beta,\sigma) = -\frac{1}{2}\left[n\log(2\pi\sigma^2) + \frac{1}{\sigma^2}\psi_n(\beta)\right].$$

Hence letting $\beta_n \in \mathcal{B}$ be such that $\psi_n(\beta_n) = \inf_{\beta\in\mathcal{B}} \psi_n(\beta)$ we get

$$(25) \qquad \ell_n(\beta_n,\sigma) = \sup_{\beta\in\mathcal{B}} \ell_n(\beta,\sigma) \qquad \text{for all } \sigma \in \mathbb{R}^+.$$

Define then $\sigma_n \in [1/u, u]$ by

$$\ell_n(\beta_n,\sigma_n) = \sup_{1/u\leqslant\sigma\leqslant u} \ell_n(\beta_n,\sigma)$$

to conclude that $\theta_n = (\beta_n,\sigma_n)$ satisfies (21).

Define now

$$(26) \quad \nu_n(\sigma) = \ell_n(\beta_n,\sigma) - \ell_n(\beta_0,\sigma) = -\frac{1}{2\sigma^2}[\psi_n(\beta_n) - \psi_n(\beta_0)].$$

By definition (21) we have $\ell_{n,\nu_0} \geqslant \ell_n(\beta_0,\sigma_n)$ and hence

$$\ell_{n,\nu} - \ell_{n,\nu_0} = \ell_n(\beta_n,\sigma_n) - \ell_{n,\nu_0} \leqslant \nu_n(\sigma_n) = \frac{\sigma_0^2}{\sigma_n^2}\, \nu_n(\sigma_0).$$

Since $\lim_{n\to\infty}\sigma_n = \sigma_0$, P_θ -a.s., we get

(27) $$\overline{\lim_{n\to\infty}} \frac{\ell_{n,\nu} - \ell_{n,\nu_0}}{\log\log n} \leqslant \overline{\lim_{n\to\infty}} \frac{\nu_n(\sigma_0)}{\log\log n}.$$

6.4. New Coordinates and Localization

As seen above, there is a finite random integer $N(\omega)$ such that P_{θ_0}-a.s., $\beta_n \in V \subset \mathcal{B}$ for $n \geqslant N(\omega)$. Hence, by (25) and the inclusion $V \subset K$

(28) $$\ell_n(\beta_n,\sigma_0) = \sup_{\beta\in\mathcal{B}} \ell_n(\beta,\sigma_0) = \sup_{\beta\in V} \ell_n(\beta,\sigma_0) \leqslant \sup_{\beta\in K} \ell_n(\beta,\sigma_0)$$

Introduce now *an arbitrary estimator* $k_n \in K = G(\mathcal{S} \times \mathcal{W})$ for β_0, such that

(29) $$\ell_n(k_n,\sigma_0) = \sup_{\beta\in K} \ell_n(\beta,\sigma_0).$$

By (26), (28) and (29) we then get trivially

(30) $$\nu_n(\sigma_0) \leqslant \ell_n(k_n,\sigma_0) - \ell_n(\beta_0,\sigma_0).$$

It is now sufficient to study the right-hand side of (30), which reduces the situation to a **local** estimation problem $k_n \in K$ with convenient local coordinates.

6.5. A "Partial" Estimation Problem

Consider *for fixed and known* $S \in \mathcal{S}$ the following partial estimation problem which will be called *the S-problem*; the parameter space is $\mathcal{W}$, convex compact neighborhood of zero in the euclidean space $\mathcal{U}$. When $U \in \mathcal{W}$ is the true value of the parameter, the observed process X is centered, stationary, gaussian with spectral density

(31) $$h_U = f_{\theta(S,U)} = \frac{\sigma_0^2}{2\pi} g_{G(S,U)} \quad \text{where } \theta(S,U) = (G(S,U),\sigma_0)$$

The standard approximate log-likelihoods $\tilde{\mathfrak{L}}_n$, $\bar{\mathfrak{L}}_n$ for the S-problem are obviously $\tilde{\mathfrak{L}}_n(U) = \tilde{L}_n[\theta(S,U)]$, $\bar{\mathfrak{L}}_n = \bar{L}_n[\theta(S,U)]$ which we write, denoting by $\mathfrak{L}_n$ the suitable one of the two functions $\tilde{\mathfrak{L}}_n$, $\bar{\mathfrak{L}}_n$,

(32) $$\mathfrak{L}_n(U) = \ell_n(G(S,U),\sigma_0) = \Lambda_n(S,U) \quad \text{for } U \in \mathcal{W}.$$

For the S-problem *we shall always assume that the true value of the parameter* U *is* 0, which implies clearly that X has distribution P_{θ_0}, since $f_{\theta(S,0)} = f_{\theta_0}$.

6.6. Uniform Consistency of the S-Problems

Consider an *arbitrary* maximum likelihood estimator $\hat{U}_n(S) \in \mathcal{W}$ for the S-problem, satisfying hence

(33) $$\mathfrak{L}_n[\hat{U}_n(S)] = \sup_{U \in \mathcal{W}} \mathfrak{L}_n(U).$$

As will become obvious below $\hat{U}_n(S)$ can be selected measurably in $S \in \mathcal{S}$. By definition (32) this is equivalent to

(34) $$\Lambda_n(S,\hat{U}_n(S)) = \sup_{U \in \mathcal{W}} \Lambda_n(S,U) \quad \text{for } S \in \mathcal{S}.$$

Moreover $\theta \to (1/n)\ell_n(\theta)$ is P_{θ_0}-a.s. an equicontinuous sequence of functions on the compact $K = G(\mathcal{S} \times \mathcal{W})$, as seen in Chapter 13, Section 2.3. Hence by (32) the family of functions $U \to (1/n)\Lambda_n(S,U)$ where n varies in $\mathbb{N}$ and S in $\mathcal{S}$ is also P_{θ_0}-a.s. equicontinuous on the compact $\mathcal{W}$.

Using this fact and (34), an argument identical to the proof of Theorem 3.2 in Chapter 13 immediately implies that P_{θ_0}-a.s. one has

$$\lim_{n\to\infty} h_{\hat{U}_n(S)} = h_0 \quad \textit{uniformly in } S \in \mathcal{S}$$

where convergence of spectral densities is understood in $\| \ \|_\infty$ on $[-\pi,\pi]$. By definition (31) of h_U, this becomes, P_{θ_0}-a.s.,

$$(35) \qquad \lim_{n\to\infty} \left[\sup_{S\in\mathcal{S}} \| g_{G(S,\hat{U}_n(S))} - g_{\beta_0} \|_\infty \right] = 0.$$

But an elementary topological argument shows that, denoting by d the euclidean distance in $\mathcal{B}$,

$$(36) \qquad \begin{cases} \text{for every } \varepsilon > 0, \text{ there is an } \eta > 0 \text{ such that the} \\ \text{constraints } \beta \in K \text{ and } \|g_\beta - g_{\beta_0}\|_\infty \leq \eta \text{ imply} \\ d(\beta,K_0) \leq \varepsilon, \text{ where } K_0 = \{\beta \in K \mid g_\beta = g_{\beta_0}\}. \end{cases}$$

Let τ be the natural projection of $\mathcal{S} \times \mathcal{W}$ onto $\mathcal{W}$. Since τG^{-1} is continuous, (35) and (36) show that P_{θ_0}-a.s. the distance between $\hat{U}_n(S) = \tau G^{-1}G(S,\hat{U}_n(S))$ and $\tau G^{-1}(K_0)$ converges to 0 *uniformly* in $S \in \mathcal{S}$.

However the definition of G implies easily the identity $K_0 = G(\mathcal{S} \times \{0\})$, and hence $\tau G^{-1}(K_0) = \{0\}$. We have thus proved that, P_{θ_0}-a.s.,

$$(37) \qquad \lim_{n\to\infty} \hat{U}_n(S) = 0 \quad \textit{uniformly in } S \in \mathcal{S}.$$

6.7. Asymptotic Information for the S-Problem

Let

$$\mathfrak{L}''_n(U) = \frac{\partial^2}{\partial U^2} \mathfrak{L}_n(U) \quad \text{and} \quad \ell''_n(\theta) = \frac{\partial^2}{\partial \theta^2} \ell_n(\theta).$$

The map $U \to \theta(S,U) = (G(S,U),\sigma_0)$ is *affine* and hence its derivative $H = (\partial/\partial U)\theta(S,U)$ is *a constant matrix*. From $\mathfrak{L}_n(U) = \ell_n[\theta(S,U)]$ we then get

$$(38) \qquad -\frac{1}{n}\mathfrak{L}''_n(U) = -\frac{1}{n} H^* \ell''_n [\theta(S,U)]H.$$

As seen in Chapter 13, Section 4.9, one has P_{θ_0}-a.s. the convergence

$$(39) \qquad \lim_{n\to\infty} -\frac{1}{n}\frac{\partial^2 \ell_n}{\partial \theta^2}(\theta) = \Phi(\theta_0,\theta) \quad \text{uniformly in } \theta \in \Theta(\nu)$$

where $\Phi(\theta_0,\theta)$ is a deterministic matrix. As shown by the explicit formula (23) in Chapter 13, $\Phi(\theta_0,\theta)$ is continuous in $\theta \in \Theta(\nu)$ and

(40) $\Phi(\theta_0,\theta) \equiv J(\theta)$ for all θ such that $f_\theta = f_{\theta_0}$

where $J(\theta)$ is the information matrix of the $\Theta(\nu)$-problem.

From (38) and (39) we deduce trivially, P_{θ_0}-a.s.,

(41) $$\lim_{n\to\infty}\left[-\frac{1}{n}\mathfrak{L}''_n(U)\right] = H^*\Phi[\theta_0,\theta(S,U)]H$$

uniformly in $(S,U) \in \mathcal{S} \times \mathcal{W}$.

Since $f_{\theta(s,0)} = f_{\theta_0}$, (40) implies, for all $S \in \mathcal{S}$,

(42) $$H^*\Phi[\theta_0,\Theta(S,0)]H = H^*J[\theta(S,0)]H.$$

We are going to prove that the matrices $\Gamma_S = H^*J[\theta(S,0)]H$ are strictly positive on the vector space $\mathcal{U}$.

As in Chapter 12, Section 3.2, we identify $\mathbb{R}^{p+q+1}$ with the natural tangent space $\mathcal{V} \times \mathbb{R}$ of $\Theta(\nu)$, where $\mathcal{V}$ is the set of pairs (A,B) of polynomials such that $d°A \leqslant p$, $d°B \leqslant q$, $A(0) = B(0) = 0$. Then $\theta(S,U)$ being *affine*, we have for all $U \in \mathcal{U}$

$$HU = \theta(S,U) - \theta(S,0) = (U,0)$$

and hence H is the natural injection of $\mathcal{U}$ in $\mathcal{V} \times \mathbb{R}$ while H^* is the natural projection of $\mathcal{V} \times \mathbb{R}$ onto $\mathcal{U}$.

On the other hand, by Chapter 12, Section 3.3, $J[\theta(S,0)]$ is a symmetric endomorphism of $\mathcal{V} \times \mathbb{R} \simeq \mathbb{R}^{p+q+1}$ whose kernel is the set of triples $(\Delta P_0, \Delta Q_0, 0)$ where Δ is any polynomial of degree $\leqslant m$, and such that $\Delta(0) = 0$. As before $m = (p - p_0) \wedge (q - q_0)$. Thus if $\underline{U} = (\overline{P},\overline{Q}) \in \mathcal{U}$ and if $HU = (\overline{P},\overline{Q},0)$ is in this kernel, we have $\overline{P} = \Delta P_0$, $\overline{Q} = \Delta Q_0$. From $d°P = d°P_0$ and $\Delta(0) = 0$ we get $\Delta \equiv 0$ and hence $U = 0$. *We have proved that* $\{\Gamma_S \mid S \in \mathcal{S}\}$ *is a (compact) set of strictly positive matrices.*

On the other hand the function

$$(S,U) \to \Gamma(S,U) = H^*\Phi[\theta_0,\theta(S,U)]H$$

is continuous on $\mathcal{S} \times \mathcal{U}$, and $\mathcal{S}$ is compact. Hence there is a compact neighborhood $\mathcal{W}_1$ of 0 in $\mathcal{U}$ such that $\Gamma(S,U)$ is strictly positive for $S \in \mathcal{S}$, $V \in \mathcal{W}_1$. From now on we shall impose on our choice of $\mathcal{W}$ (cf. Section 5.3) the restriction $\mathcal{W} \subset \mathcal{W}_1$. Consequently

(43) $\Gamma(\mathcal{S} \times \mathcal{W})$ is a compact set of strictly positive matrices.

By (41) we may then asert the existence of a *finite* random integer $N_1(\omega) \geqslant N(\omega)$ such tht P_{θ_0}-a.s.

(44) $$-\frac{1}{n}\ \mathfrak{L}''_n(U) > 0 \quad \text{for all} \quad n \geqslant N_1(\omega),\ \ S \in \mathcal{S},\ \ U \in \mathcal{W}.$$

6.8. Link Between Global and Partial Maximal Likelihood

By (44), for $n \geqslant N_1(\omega)$, and $S \in \mathcal{S}$, the function $\mathfrak{L}_n$ is strictly concave on $\mathcal{W}$, and hence the estimator $\hat{U}_n(S)$ is *uniquely* determined by (33). Since $\Lambda_n(S,U)$ is continuous, (34) now shows, by a standard compactness argument, at fixed n, that *the function* $S \to \hat{U}_n(S)$ *must be continuous for* $n \geqslant N_1(\omega)$.

We may then find an estimator $S_n \in \mathcal{S}$ such that for $n \geqslant N_1(\omega)$

(45) $$\Lambda_n(S_n,\hat{U}_n(S_n)) = \sup_{S\in\mathcal{S}}\ \Lambda_n(S,\hat{U}_n(S)).$$

Setting $k_n = G(S_n,\ \hat{U}_n(S_n)) \in K$, we now get from definitions (32) and (45)

$$\ell_n(k_n,\sigma_0) = \Lambda_n(S_n,\hat{U}_n(S_n)) = \sup_{S\in\mathcal{S}}\ \sup_{U\in\mathcal{W}}\ \Lambda_n(S,U) = \sup_{\beta\in K}\ \ell_n(\beta,\sigma_0).$$

Thus for $n \geqslant N_1(\omega)$, the estimator k_n satisfies (29) and as seen in 6.4 this implies, by (30),

(46) $$v_n(\sigma_0) \leqslant \ell_n(k_n,\sigma_0) - \ell_n(\beta_0,\sigma_0) = w_n(S_n) \qquad \text{for } n \geqslant N_1(\omega)$$

where we have **defined** $w_n(S)$ by

(47) $$w_n(S) = \sup_{U\in\mathcal{W}}\ \Lambda_n(S,U) - \Lambda_n(S,0) = \mathfrak{L}_n[\hat{U}_n(S)] - \mathfrak{L}_n(0).$$

6.9. Asymptotic Expression of $w_n(S)$

Recall that the $\hat{U}_n(S)$ are uniformly consistent by (37) and that $\mathcal{W}$ is a neighborhood of 0. Hence there is a P_{θ_0}-a.s. finite random integer $N_2 \geqslant N_1$ such that

(48) $n \geqslant N_2(\omega)$ implies that $\hat{U}_n(S)$ is interior to $\mathcal{W}$ for all $S \in \mathcal{S}$.

Denote by $\mathfrak{L}'_n$, $\mathfrak{L}''_n$ the first two partial derivatives of $\mathfrak{L}_n$ with respect to U. For $n \geqslant N_2(\omega)$ we must have $\mathfrak{L}'_n[\hat{U}_n(S)] \equiv 0$ for all $S \in \mathcal{S}$. Taylor's formula used twice then implies

(49) $$0 = \mathfrak{L}_n'(\hat{U}_n) = \mathfrak{L}_n'(0) + \mathfrak{L}_n''(\mu_n\hat{U}_n)\cdot\hat{U}_n$$

(50) $$\mathfrak{L}_n(0) = \mathfrak{L}_n(\hat{U}_n) + \frac{1}{2}\hat{U}_n^*\mathfrak{L}_n''(\tau_n\hat{U}_n)\hat{U}_n$$

where $\mu_n(S)$, $\tau_n(S)$ are suitable random numbers in $[0,1]$.
Introduce the notation

(51) $$Y_n = \frac{1}{\sqrt{n}}\mathfrak{L}_n'(0); \quad M_n = -\frac{1}{n}\mathfrak{L}_n''(\mu_n\hat{U}_n); \quad K_n = -\frac{1}{n}\mathfrak{L}_n''(\tau_n U_n).$$

We point out that $Y_n(S)$, $M_n(S)$, $K_n(S)$ are clearly jointly measurable functions of $X_1 \dots X_n$ and $S \in \mathcal{S}$. From (44) we deduce that P_{θ_0}-a.s.

(52) all the $M_n(S)$, $K_n(S)$ are invertible for $S \in \mathcal{S}$, $n \geqslant N_2(\omega)$.

For $n \geqslant N_2(\omega)$ we first get, from (49), $\hat{U}_n = (1/\sqrt{n})M_n^{-1}Y_n$ and then from (50) and (47), P_{θ_0}-a.s.,

(53) $$w_n(S) = \mathfrak{L}_n[\hat{U}_n(S)] - \mathfrak{L}_n(0) = \frac{1}{2}(Y_m^* M_n^{-1}K_n M_n^{-1}Y_n)$$

$$\text{for all } n \geqslant N_2(\omega), \quad S \in \mathcal{S}.$$

Since $|\mu_n| \leqslant 1$ and $|\tau_n| \leqslant 1$, the uniform consistency of the $\hat{U}_n(S)$ implies

(54) $$\lim_{n\to\infty} \sup_{S\in\mathcal{S}} |\mu_n\hat{U}_n| = \lim_{n\to\infty} \sup_{S\in\mathcal{S}} |\tau_n\hat{U}_n| = 0, \quad P_{\theta_0}\text{-a.s.}$$

From the uniform convergence (41) of $(-(1/n)\mathfrak{L}_n'')$ and (54) it follows readily that P_{θ_0}-a.s.

(55) $$\lim_{n\to\infty} M_n(S) = \lim_{n\to\infty} K_n(S) = H^*\Phi[\theta_0, \theta(S,0)]H$$

uniformly in $S \in \mathcal{S}$

and the right-hand side is (cf. (42)) the strictly positive matrix Γ_S.

By an elementary deterministic argument, (55) proves the existence of a fixed sequence of numerical random variables $\varepsilon_n > 0$ such that $\lim_{n\to\infty} \varepsilon_n = 0$, P_{θ_0}-a.s., and

$$M_n(S)^{-1}K_n(S)M_n(S)^{-1} \leqslant (1 + \varepsilon_n)\Gamma_S^{-1} \text{ for all } n \geqslant N_2(\omega), \text{ all } S \in \mathcal{S}.$$

Identifying gradients with column vectors and using only definitions (32) and $\ell'_n(\theta) = (\partial/\partial\theta)\ell_n(\theta)$ we get

$$\mathfrak{L}'_n(U) = \frac{\partial}{\partial U}\,\ell_n[\theta(S,U)] = H^*\ell^*_n[\theta(S,U)]$$

whence

$$(57)\qquad Y_n = \frac{1}{\sqrt{n}}\,\mathfrak{L}'_n(0) = \frac{1}{\sqrt{n}}\,H^*\ell'_n[\theta(S,0)].$$

Remembering that ℓ_n is either $\bar{L}_n$ or $\check{L}_n$ we call $\bar{Y}_n$, $\check{Y}_n$ the respective values of Y_n when $\ell_n = \bar{L}_n$ and $\ell_n = \check{L}_n$.

Consider first the case $\ell_n = \bar{L}_n$. As seen in Chapter 13, Section 4.10 we have for all $\theta \in \Theta(\nu)$

$$(58)\qquad \frac{1}{\sqrt{n}}\frac{\partial \bar{L}_n}{\partial\theta}(\theta) = \sqrt{n}[I_n(\varphi) - I(\varphi)] \quad\text{with}\quad \varphi = \frac{1}{4\pi f_\theta^2}\frac{\partial f_\theta}{\partial\theta}$$

where I_n is the periodogram of X and $I(\varphi) = \int_{\mathbb{T}} \varphi f_\theta d\lambda$. For $\theta = \theta(S,0) = (P_0S,\ Q_0S,\ \sigma_0)$ we have $f_{\theta(S,0)} = f_{\theta_0}$ and we define a family $(\varphi_S)_{S\in\mathcal{S}}$ of functions φ_S: $\mathbb{T} \to \mathbb{R}$ by

$$(59)\qquad \varphi_S = \frac{1}{4\pi f_{\theta_0}^2}\,H^*\,\frac{\partial f_\theta}{\partial\theta}(P_0S,\ Q_0S,\ \sigma_0).$$

Then in the case $\ell_n = \bar{L}_n$, (57) and (58) give

$$(60)\qquad Y_n = \bar{Y}_n = \sqrt{n}[I_n(\varphi_S) - I(\varphi_S)].$$

To study the case $\ell_n = \check{L}_n$ recall that by Chapter 13, Sections 4.6 and 4.7,

$$\sup_{\theta\in\Theta(\nu)}\left|\frac{\partial \bar{L}_n}{\partial\theta} - \frac{\partial \check{L}_n}{\partial\theta}\right|$$

is uniformly bounded in $L^4(\Omega, P_{\theta_0})$ where n varies in $\mathbb{N}$.

Hence (57) shows that for some deterministic constant c

$$E_{\theta_0}\,[\sup_{S\in\mathcal{S}} |\bar{Y}_n - \check{Y}_n|^4] \leq \frac{c}{n^2} \qquad \text{for all } n \geq 0$$

which implies

$$P_{\theta_0}\left[\sup_{S\in\mathcal{S}} |\bar{Y}_n - \check{Y}_n| \geq \frac{1}{n^{1/5}}\right] \leq \frac{c}{n^{6/5}}$$

and by Borel-Cantelli's lemma, there is a finite random

integer $N_3 \geq N_2$ such that P_{θ_0}-a.s.

$$(61) \qquad \sup_{S \in \mathcal{S}} |\bar{Y}_n - \tilde{Y}_n| \leq \frac{1}{n^{1/5}} \quad \text{for all } n \geq N_3(\omega).$$

The norms $\|\Gamma_S^{-1}\|$ remain bounded by a constant $c > 0$ when $S \in \mathcal{S}$. Expanding the quadratic form $\tilde{Y}_n^* \Gamma_S^{-1} \tilde{Y}_n$ we then get from (61), P_{θ_0}-a.s.,

$$\tilde{Y}_n^* \Gamma_S^{-1} \tilde{Y}_n \leq \bar{Y}_n \Gamma_S^{-1} \bar{Y}_n + \frac{2c}{n^{1/5}} |\bar{Y}_n| + \frac{c}{n^{2/5}} \quad \text{for all } n \geq N_3(\omega),\ S \in \mathcal{S}.$$

But there is another constant c such that for all vectors $v \in \mathbb{R}^{\dim \mathcal{U}}$, all $S \in \mathcal{S}$,

$$|v| \leq c(v^* \Gamma_S^{-1} v)^{1/2} \leq c(1 + v^* \Gamma_S^{-1} v)$$

whence finally, P_{θ_0}-a.s.

$$(62) \qquad \tilde{Y}_n^* \Gamma_S^{-1} \tilde{Y}_n \leq \left[1 + \frac{c}{n^{1/5}}\right] \bar{Y}_n^* \Gamma_S^{-1} \bar{Y}_n + \frac{c}{n^{1/5}}$$

for all $n \geq N_3(\omega)$ all $S \in \mathcal{S}$. Combining (56) and (62) we see that whether $\ell_n = L_n$ or $\ell_n = L_n$, there is a sequence of numerical random variables $\eta_n > 0$ converging P_{θ_0}-a.s. to zero, such that one has P_{θ_0}-a.s.,

$$(63) \quad w_n(S) \leq \frac{1}{2}(1 + \eta_n)(\bar{Y}_n \Gamma_S^{-1} \bar{Y}_n) + \eta_n \text{ for all } n \geq N_3(\omega), \textit{all } S \in \mathcal{S}$$

6.10. Final estimate of $(\ell_{n,\nu} - \ell_{n,\nu_0})$

We now replace S by S_n in (63) and gathering (27), (46) and (63) we get the chain of inequalities

$$(64) \qquad \begin{aligned} \overline{\lim_{n\to\infty}} \frac{\ell_{n,\nu} - \ell_{n,\nu_0}}{\log \log n} &\leq \overline{\lim_{n\to\infty}} \frac{v_n(\sigma_0)}{\log \log n} \\ &\leq \overline{\lim_{n\to\infty}} \frac{w_n(S_n)}{\log \log n} \leq \overline{\lim_{n\to\infty}} \left[\sup_{S \in \mathcal{S}} \frac{\bar{Y}_n \Gamma_S^{-1} \bar{Y}_n}{2 \log \log n}\right]. \end{aligned}$$

Define $F_n^X(g) = (n/2 \log \log n)^{1/2}[I_n(g) - I(g)]$ and

$$\Gamma(g) = 4\pi \int_{\mathbb{T}} g g^* f_{\theta_0}^2 \, d\lambda.$$

In view of the expression (16), Chapter 13 for $J(\theta)$ and of definition (59) for φ_S, we have the relation

$$\Gamma_S = H^* J[\theta(S,0)]H = \Gamma(\varphi_S)$$

and finally

$$\frac{\overline{Y}_n^* \Gamma_S^{-1} \overline{Y}_n}{2 \log \log n} = F_n^X(\varphi_S)^* \Gamma(\varphi_S)^{-1} F_n^X(\varphi_S).$$

Call $\mathcal{F}$ the compact set of functions φ_S, $S \in \mathcal{S}$.

We now want to apply *the crucial Theorem* 7.5 *below* to get, P_{θ_0}-a.s.,

$$\overline{\lim_{n\to\infty}} \left[\sup_{g\in \mathcal{F}} F_n^X(g)^* \Gamma(g)^{-1} F_n^X(g) \right] \leq p + q - m \tag{65}$$

where $p + q - m = p + q - (p - p_0) \wedge (q - q_0)$ is the dimension of the matrices $\Gamma(g)$. The only hypothesis to check is that the functions $g \in \mathcal{F}$ have Fourier coefficients decreasing at *uniform* exponential speed. But this is implied by Lemma 7.2 below. This concludes the proof of Theorem 4.8, since (64) and (65) yield P_{θ_0}-a.s.

$$\overline{\lim_{n\to\infty}} \frac{\ell_{n,\nu} - \ell_{n,\nu_0}}{\log \log n} \leq p + q - (p - p_0) \wedge (q - q_0).$$

7. Law of the Iterated Logarithm for the Periodogram

7.1. Exponentially Decreasing Fourier Coefficients

Call $C(\mathbb{T})$ the set of continuous functions g: $\mathbb{T} \to \mathbb{R}$ with $\mathbb{T} = [-\pi,\pi]$, endowed with the $\| \ \|_\infty$-norm, and call

$$\hat{g}_k = \frac{1}{2\pi} \int_{-\pi}^{\pi} g(\lambda) e^{ik\lambda} d\lambda$$

the Fourier coefficients of g.

For $0 < \beta < 1$ and $B > 0$ we define an obviously compact subset $H(\beta,B)$ of $C(\mathbb{T})$ by

$$H(\beta,B) = \{g \in C(\mathbb{T}) | g \text{ is } \textit{even} \text{ and } |\hat{g}_k| \leq B\, \beta^{|k|} \text{ for all } k \in \mathbb{Z}\}. \tag{66}$$

When g: $\mathbb{T} \to \mathbb{R}^r$ is *vector valued*, we shall still say that g *is of type* $H(\beta,B)$ if all its coordinates belong to $H(\beta,B)$.

For $g \in H(\beta,B)$ we have $g(\lambda) = \Sigma_j g_j e^{-ij\lambda}$ and hence

(67) $$\|g\|_\infty \leqslant \frac{2B}{1-\beta} .$$

For $g,h \in H(\beta,B)$, the classical formula holds

(68) $$(gh)_j = \sum_k \hat{g}_{j-k}\hat{h}_k$$

and proves readily the existence of β', B' depending only on β, B such that

(69) $$g \text{ and } h \text{ in } H(\beta,B) \text{ implies } gh \in H(\beta',B')$$

The sets $H(\beta,B)$ arise naturally in the context of ARMA processes, as shown by the following result.

7.2. Lemma. *Let $\Theta = \Theta(\nu,\rho,u)$ be any one of the parameter spaces introduced in* 2.1. *For $\theta = (P,Q,\sigma) \in \Theta$ denote by*

$$f_\theta(\lambda) = \frac{\sigma^2}{2\pi}\left|\frac{Q}{P}(e^{-i\lambda})\right|^2$$

the corresponding spectral density. Then there are numbers α, A with $0 < \alpha < 1$, $A > 0$ such that for all $\theta,\theta_0 \in \Theta$ the functions f_θ, $\partial f_\theta/\partial\theta$, $1/f^2_{\theta_0} \times \partial f_\theta/\partial\theta$ are of type $H(\alpha,A)$.

Proof. The definition of f_θ shows that the functions $f_\theta(\lambda)$, $1/f_\theta(\lambda)$, and all the coordinates of $\partial f_\theta/\partial\theta(\lambda)$ are of the form $\varphi_\theta(\lambda) = \psi_\theta(e^{-i\lambda})$ where the functions $\psi_\theta(z)$ are rational fractions in $z \in \mathbb{C}$ with coefficients continuous in $\theta \in \Theta$ and *no poles in the domain* $1/1+\rho < |z| < 1 + \rho$. Fix r with $0 < r < \rho$ and call D, D_r the circles of center 0 and radiuses 1, $1 + r$, oriented in the usual trigonometric sense. We have, setting $z = e^{-i\lambda}$

$$(\hat{\varphi}_\theta)_k = \frac{1}{2\pi}\int \varphi_\theta(\lambda)e^{ik\lambda}d\lambda = \frac{1}{2\pi i}\int_D z^{-k-1}\psi_\theta(z)dz$$

and by Cauchy's formula

$$(\hat{\varphi}_\theta)_k = \frac{1}{2\pi i}\int_{D_r} z^{-k-1}\psi_\theta(z)dz.$$

Since

$$\sup_{\theta\in\Theta}\ \sup_{|z|=1+r} |\psi_\theta(z)|$$

is a finite constant c we get

$$|(\hat{\varphi}_\theta)_k| \leq \frac{c}{(1+r)^k}$$

for all $k \geq 0$ all $\theta \in \Theta$. Hence all the (obviously even) functions f_θ, $1/f_\theta$, $\partial f_\theta/\partial\theta$ lie in some fixed $H(\alpha,A)$. By (69) the

$$\frac{1}{f_{\theta_0}^2}\frac{\partial f_\theta}{\partial\theta}$$

being the product of three functions of type $H(\alpha,A)$ must also lie in some fixed $H(\alpha',A')$ when θ_0, θ vary in Θ.

7.3. Theorem. *Let X be a centered gaussian stationary* ARMA *process with spectral density f. Call P_f the global law of X. Consider on $C(\mathbb{T})$ the random functional $g \to I_n^X(g)$ defined* (cf. *Chapter* 10) *by the periodogram of X and its almost sure deterministic limit*

$$I^X(g) = \int_{\mathbb{T}} fg\, d\lambda.$$

Fix any compact set $H(\beta,B)$ of even functions in $C(\mathbb{T})$ as in 7.1. *Then P_f*-a.s. *one has*

$$\overline{\lim_{n\to\infty}}\ \frac{n}{2\log\log n}[I_n^X(g) - I^X(g)] = \left[4\pi\int_{\mathbb{T}} f^2g^2d\lambda\right]^{1/2}$$

uniformly in $g \in H(\beta,B)$.

Proof. The proof is rather intricate and will only be completed in Section 7.11, after reduction to the case of gaussian white noise. We point out that *when g is odd $I_n^X(g) = I^X(g) = 0$.* This explains the restriction of Theorem 7.3 to even functions.

Note also that $(4\pi \int_{\mathbb{T}} f^2g^2d\lambda)$ *is the asymptotic variance of* $\sqrt{n}[I_n^X(g) - I^X(g)]$ as seen in Chapter 10, Theorem 3.1.

7.4. Generalization

The same result and the same proof hold if X is only assumed to be of the form $X_n = \Sigma_{k\geq 0}u_kW_{n-k}$ where W is a gaussian white noise, provided there is an $\alpha < 1$ and an $A > 0$ such that

$|u_k| \leqslant A\alpha^k$ for all $k \geqslant 0$.

7.5. Theorem (multidimensional version). *Let X be a centered gaussian* ARMA *process with density f as in Theorem* 7.3. *Fix a compact set $\mathcal{F}$ of vector valued,* **even,** *continuous functions g: $\mathbb{T} \to \mathbb{R}^r$. Define as before*

$$F_n^X(g) = (n/2 \log \log n)^{1/2}[I_n^X(g) - I^X(g)].$$

Assume that for some fixed $\beta < 1$ and $B > 0$, all the $g \in \mathcal{F}$ are of type $H(\beta,B)$ - cf. Definition 7.1. *Assume that for each $g \in \mathcal{F}$ the matrix $\Gamma(g) = 4\pi\int_{\mathbb{T}} gg^* f^2 d\lambda$ is invertible. Then the r-dimensional vectors $F_n^X(g)$ satisfy the asymptotic inequality*

$$\overline{\lim_{n\to\infty}} \left[\sup_{g\in\mathcal{F}} F_n^X(g)^*\Gamma(g)^{-1}F_n^X(g)\right] \leqslant r, \qquad P_f\text{-a.s.}$$

Proof. Let $h = \Gamma(g)^{-1/2}g$ with $g \in \mathcal{F}$. When g varies in $\mathcal{F}$, h remains of fixed type $H(\beta,B_1)$ for some $B_1 > 0$. Call h_j, $1 \leqslant j \leqslant r$, the coordinates of h. We have

$$F_n^X(g)^*\Gamma(g)^{-1}F_n^X(g) = \|F_n^X(h)\|^2 = \sum_{j=1}^{r} F_n^X(h_j)^2.$$

Then by Theorem 7.3, P_f-a.s. we have uniformly in $h \in H(\beta,B_1)$

$$\overline{\lim_{n\to\infty}} F_n^X(h_j)^2 = 4\pi \int_{\mathbb{T}} h_j^2 f^2 d\lambda, \quad 1 \leqslant j \leqslant r.$$

The last two relations give immediately, P_f-as.

$$\overline{\lim_{n\to\infty}} F_n^X(g)^*\Gamma(g)^{-1}F_n^X(g) \leqslant 4\pi \int_{\mathbb{T}} |h|^2 f^2 d\lambda \tag{70}$$

uniformly in $g \in \mathcal{F}$. But by definition, using trace M_1M_2 = trace M_2M_1, we have

$$|h|^2 = g^*\Gamma(g)^{-1}g = \text{trace } g^*\Gamma(g)^{-1}g = \text{trace}(\Gamma(g)^{-1}gg^*)$$

and hence denoting by Id_r the identity matrix in dimension r,

$$4\pi\int_{\mathbb{T}} |h|^2 f^2 d\lambda = \text{trace}\left[4\pi\Gamma(g)^{-1}\int_{\mathbb{T}} gg^* f^2 d\lambda\right] = \text{trace}(Id_r) = r$$

which achieves the proof.

7.6. The Notations $O(\cdot)$ and $O_{a.s.}(\cdot)$

Let u_n be any deterministic sequence of positive numbers and let V_n be any sequence of random variables. We write $V_n = O(u_n)$ iff there is a *deterministic constant* c such that $|V_n| \leqslant cu_n$ for all $n \geqslant 1$, P_f-a.s.

We write $V_n = O_{a.s.}(u_n)$ iff there is a positive *random variable C which is P_f-a.s. finite* and such that $|V_n| \leqslant Cu_n$ for all $n \geqslant 1$, P_f-a.s.

Recall the following elementary result, valid for any fixed $r > 0$, $\varepsilon > 0$:

(71) if $\|V_n\|_r = O(u_n)$ then one has $V_n = O_{a.s.}(n^{(1/r)+\varepsilon}\, u_n)$.

Indeed the series

$$\sum_{n\geqslant 0} P_f(|V_n| > n^{(1/r)+\varepsilon}\, u_n)$$

is then bounded by

$$c \sum_{n\geqslant 0} \frac{1}{n^{1+r}} < \infty$$

and Borel-Cantelli's lemma proves (71).

7.7. Reduction to the White Noise Case

Lemma. *Assume that W is a gaussian white noise on a probability space (Ω,P) and that*

(72) $\qquad X_n + \sum_{r\geqslant 0} c_r W_{n-r} \qquad$ *with* $\quad |c_r| \leqslant A\alpha^r$

where $\alpha < 1$ and $A > 0$ are fixed. Call f the spectral density of X and σ^2 the variance of W. Call I_n^X, I_n^W the periodograms of X, W and I^X, I^W the associated deterministic functionals. Set

(73) $\qquad F_n^X(g) = (n/2\log\log n)^{1/2}[I_n^X(g) - I^X(g)]$

with a similar definition for $F_n^W(g)$. Fix a compact set $H(\beta,B) \subset C(\mathbb{T})$ as in 7.1. Then one has, P-a.s.,

(74) $$\lim_{n\to\infty} \sup_{g\in H(\beta,B)} \left| F_n^X(g) - F_n^W\left(\frac{2\pi f}{\sigma^2}\, g\right)\right| = 0.$$

Proof. Call $\hat{r}_j(n)$ and $\hat{w}_j(n)$ the estimated autocovariances of X and W, given by

$$(75)\qquad \hat{r}_j(n) = \hat{r}_{-j}(n) = \frac{1}{n}\sum_{k=1}^{n-j} X_k X_{k+j} \quad \text{for } 0 \leqslant j \leqslant n-1$$

$$(76)\qquad \hat{w}_j(n) = \hat{w}_{-j}(n) = \frac{1}{n}\sum_{k=1}^{n-j} W_k W_{k+j} \quad \text{for } 0 \leqslant j \leqslant n-1$$

and set $\hat{r}_j(n) = \hat{w}_j(n) = 0$ for $|j| \geqslant n$. The definitions of Chapter 10, Section 2.1 yield foir $g \in H(\text{ß},B)$ with Fourier coefficients $\hat{g}_j$

$$(77)\qquad I_n^X(g) = \sum_j \hat{r}_j(n)\hat{g}_j \quad \text{and} \quad I^X(g) = \int_{\mathbb{T}} fg\, d\lambda$$

$$(78)\qquad I_n^W(g) = \sum_j \hat{w}_j(n)\hat{g}_j \quad \text{and} \quad I^W(g) = \frac{\sigma^2}{2\pi}\int_{\mathbb{T}} g\, d\lambda = \sigma^2\hat{g}_0.$$

Since $I^X(g) \equiv I^W\left[\frac{2\pi}{\sigma^2} fg\right]$, we have

$$(79)\qquad F_n^X(g) - F_n^W\left(\frac{2\pi}{\sigma^2} fg\right)$$

$$= (n/2\log\log n)^{1/2}\left[I_n^X(g) - I_n^w\left(\frac{2\pi}{\sigma^2} fg\right)\right].$$

With no loss of generality we may of course assume $\sigma^2 = 1$ and hence we now have to study

$$(80)\qquad \Phi_n(g) = I_n^X(g) - I_n^W(2\pi fg).$$

The Fourier coefficients of $2\pi fg$ are obviously

$$(2\pi fg)_j = \sum_u 2\pi\hat{f}_{j-u}\hat{g}_u = \sum_u r_{j-u}\hat{g}_u$$

where the r_k are the covariances of X, so that (78) implies

$$(81)\qquad I_n^W(2\pi fg) = \sum_{j,u} \hat{w}_j(n) r_{j-u}\hat{g}_u = \sum_u s_u(n)\hat{g}_u$$

with the notation

$$(82)\qquad s_u(n) = \sum_j r_{j-u}\hat{w}_j(n) = \sum_v r_v \hat{w}_{u-v}(n).$$

The expansion (72) of X in terms of W allows us to compute the covariances of X

(83) $$r_j = E[X_0 X_{|j|}] = \sum_{r \geqslant 0} c_r c_{r+|j|} \quad \text{for } j \in \mathbb{Z}$$

which forces

(84) $$|r_j| \leqslant \frac{A^2}{(1-\alpha)^2} \alpha^{|j|} \quad \text{for } j \in \mathbb{Z}.$$

Consequently for any $u \geqslant 0$ we get from (82) the elementary algebraic identity

(85) $$\begin{aligned} s_u(n) &= \sum_v r_v \hat{w}_{u-v}(n) = \sum_v \sum_{r \geqslant 0} c_r c_{r+|v|} \hat{w}_{u-v}(n) \\ &= \sum_{r \geqslant 0} \sum_{s \geqslant 0} c_r c_s \hat{w}_{u+r-s}(n) \end{aligned}$$

and obviously $s_u(n) \equiv s_{-u}(n)$.

On the other hand, by simple reshuffling of sums, (72) and (75) imply

(86) $$\hat{r}_j(n) = \sum_{r \geqslant 0} \sum_{s \geqslant 0} c_r c_s F_{rs}(j,n) \quad \text{for } |j| \leqslant n-1$$

with the definition

(87) $$F_{rs}(j,n) = \frac{1}{n} \sum_{k=1}^{n-|j|} W_{k-r} W_{k+|j|-s} \quad \text{for } |j| \leqslant n-1,\ 0 \leqslant r,s.$$

From (85) and (86) we now obtain for $|j| \leqslant n-1$

(88) $$\hat{r}_j(n) - s_j(n) = \sum_{r \geqslant 0} \sum_{s \geqslant 0} c_r c_s [F_{rs}(j,n) - \hat{w}_{|j|+r-s}(n)]$$

while on the other hand by (77), (80), and (81) we get

(89) $$\Phi_n(g) = \sum_j [\hat{r}_j(n) - s_j(n)] \hat{g}_j.$$

Define now

(90) $$G_{rs}(j,n) = F_{rs}(j,n) - \hat{w}_{r+|j|-s}(n) \text{ for } r,s \geqslant 0,\ |j| \leqslant n-1.$$

An elementary check based only on definitions (76) and (87) shows that $[nG_{rs}(j,n)]$ is the sum of a finite number $M(r\ s\ j\ n)$ of random variables of the form $(\pm W_m W_{m'})$ with

$$M(r\ s\ j\ n) = r + s \quad \text{for} \quad |r + |j| - s| \leq n - 1$$

$$M(r\ s\ j\ n) = n - |j| \quad \text{for} \quad |r + |j| - s| \geq n.$$

In particular for $0 \leq r,s < b \log n$ and $|j| \leq n-1$ we always have

$$M(r\ s\ j\ n) \leq 2b \log n.$$

Hence the set E_n of all W_m, $W_{m'}$ involved in writing down all the expressions $nG_{rs}(j,n)$, where $0 \leq r,s < b \log n$ and $|j| \leq n-1$, has cardinal

$$|E_n| \leq 2(2b \log n)(b \log n)^2(2n) = 8b^3 n(\log n)^3.$$

Since W_t is standard gaussian, one has

$$(91) \qquad P(|W_t| > 3\sqrt{\log n}\,) \leq e^{-(9/2)\log n} = \frac{1}{n^{9/2}}$$

for all t, n and hence

$$(92) \qquad P\left[\sup_{t \in E_n} |W_t| > 3\sqrt{\log n}\right] \leq |E_n| P(|W_t| > 3\sqrt{\log n}\,) \leq \frac{8b^3(\log n)^3}{n^{7/2}}.$$

Thus by Borel-Cantelli's lemma, there is a finite random integer $N(\omega)$ such that P-a.s., $n \geq N(\omega)$ implies

$$\left\{ \sup_{t \in E_n} |W_t| \leq 3\sqrt{\log n} \right\}.$$

A fortiori, P-a.s., we get for all $0 \leq r,s < b \log n$, all $|j| \leq n-1$, all $n \geq N(\omega)$, the bound

$$(93) \qquad |nG_{rs}(j,n)| \leq M(r\ s\ j\ n)(3\sqrt{\log n}\,)^2 \leq 18b(\log n)^2.$$

This implies, using (71) and the definition of $G_{rs}(j,n)$,

$$(94) \qquad \sup_{0 \leq r,s < b \log n} \ \sup_{|j| \leq n-1} |F_{rs}(j,n) - \hat{w}_{j+|j|-s}(n)| = O_{a.s.}\left[\frac{(\log n)^2}{n}\right]$$

On the other hand, by definitions (76) and (87) we obviously have

$$(95) \qquad \|F_{rs}(j,n)\|_2 \leq 1, \ \|\hat{w}_{r+|j|-s}(n)\|_2 \leq 1 \ \text{for} \ |j| \leq n-1,\ 0 \leq r,s$$

and provided $b > 3\log(1/\alpha)$,

$$\sum_{(r\vee s)\geqslant b\log n} |c_r\, c_s| \leqslant \frac{2A\ \alpha^{b\log n}}{(1-\alpha)^2} = O(1/n^3).$$

Consequently for $|j| \leqslant n-1$, we obtain

$$\Big\| \sum_{(r\vee s)\geqslant b\log n} c_r c_s F_{rs}(j,n) \Big\|_2 = O(1/n^3).$$

The obvious inequality $\sup_j |V_j| \leqslant \sum_j |V_j|$ yields then

$$\Big\| \sup_{|j|\leqslant n-1} \Big| \sum_{(r\vee s)\geqslant b\log n} c_r c_s F_{rs}(j,n) \Big| \Big\|_2 = O(1/n^2)$$

and hence by (71)

$$\sup_{|j|\leqslant n-1} \Big| \sum_{(r\vee s)\geqslant b\log n} c_r c_s F_{rs}(j,n) \Big| = O_{\text{a.s.}}(1/n). \tag{96}$$

An *identical* argument implies

$$\sup_{|j|\leqslant n-1} \Big| \sum_{(r\vee s)\geqslant b\log n} c_r c_s \hat{w}_{r+|j|-s}(n) \Big| = O_{\text{a.s.}}(1/n). \tag{97}$$

From (88), (94), (96) and (97) we immediately obtain

$$\sup_{|j|\leqslant n-1} |\hat{r}_j(n) - s_j(n)| = O_{\text{a.s.}}\left[\frac{(\log n)^4}{n}\right]. \tag{98}$$

On the other hand since $\hat{r}_j(n) = 0$ for $|j| \geqslant n$, we have, by Cauchy-Schwartz,

$$\Big| \sum_{|j|\geqslant n} [\hat{r}_j(n) - s_j(n)]\hat{g}_j \Big| \leqslant \Big[\sum_{|j|\geqslant n} |s_j(n)|^2 \Big]^{1/2} \Big[\sum_{|j|\geqslant n} |\hat{g}_j|^2 \Big]^{1/2}. \tag{99}$$

Since $\hat{w}_j(n) = 0$ for $|j| \geqslant n$, the definition (82) of $s_u(n)$ gives for $u \geqslant 0$

$$\|s_u(n)\|_2 = \Big\|\sum_v r_v \hat{w}_{u-v}(n)\Big\|_2 = \Big\| \sum_{v\geqslant u-n+1} r_v \hat{w}_{u-v}(n) \Big\|_2$$

$$\leqslant \sum_{v\geqslant u-n+1} |r_v| \leqslant \frac{A^2}{(1-\alpha)^3}\alpha^{u-n+1}$$

and consequently

$$E\Big[\sum_{u\geqslant n-1} |s_u(n)|^2 \Big] = \sum_{u\geqslant n-1} \|s_u(n)\|_2^2 = O(1).$$

In view of (71), this implies for any fixed $\varepsilon > 0$

$$\sum_{u \geqslant n-1} |s_u(n)|^2 = O_{a.s.}(n^{1+\varepsilon}).$$

By definition of $H(\beta,B)$, one has

$$\sup_{g \in H(\beta,B)} \left(\sum_{j \geqslant n} |\hat{g}_j|^2 \right) \leqslant \frac{B^2}{1-\beta^2} \beta^{2n}$$

and hence (99) now implies

$$\sup_{g \in H(\beta,B)} \left| \sum_{|j| \geqslant n} [\hat{r}_j(n) - s_j(n)]\hat{g}_j \right| = O_{a.s.}(n\beta^n) = O_{a.s.}(1/n). \tag{100}$$

We bound the remaining terms in the series $\Phi_n(g)$ with the help of (67) and (98) to get

$$\sup_{g \in H(\beta,B)} \left| \sum_{|j| \leqslant n-1} [\hat{r}_j(n) - s_j(n)]\hat{g}_j \right| \leqslant \left[\sup_{g \in H(\beta,B)} \|g\|_\infty \right] \times O_{a.s.}\left[\frac{(\log n)^4}{n}\right] = O_{a.s.}\left[\frac{(\log n)^4}{n}\right]. \tag{101}$$

Combining (101), (100) and (89) we obtain

$$\sup_{g \in H(\beta,B)} |\Phi_n(g)| = O_{a.s.}\left[\frac{(\log n)^4}{n}\right].$$

By (79) and (80) we finally get, taking into account the fact that σ^2 had been taken equal to 1 in the main part of the proof,

$$\sup_{g \in H(\beta,B)} \left| F_n^X(g) - F_n^W\left[\frac{2\pi}{\sigma^2} fg\right] \right| = O_{a.s.}\left[\frac{(\log n)^4}{\sqrt{n \log \log n}}\right]$$

which proves Lemma 7.7.

7.8. *Simultaneous Bounds for Estimated Autocovariances of White Noise*

Lemma. *Let w be a centered gaussian white noise of variance 1 with global law P. Let $\hat{w}_j(n)$ be the estimated covariances of W based on observations $W_1 \dots W_n$, given by (76). Then for any constant b one has P-almost surely*

$$\overline{\lim_{n\to\infty}} \left[(n/2\log \log n)^{1/2} \, |\hat{w}_0(n) - 1| \right] = \sqrt{2} \tag{102}$$

(103) $$\overline{\lim_{n\to\infty}} \left[(n/2\log\log n)^{1/2} \max_{1 \leqslant j \leqslant b\log n} |\hat{w}_j(n)| \right] \leqslant \sqrt{2}\,.$$

Proof. Define

(104) $$S_0(n) = \sum_{k=1}^{n} (W_k^2 - 1) = n(\hat{w}_0(n) - 1)$$

(105) $$S_{-j}(n) = S_j(n) = \sum_{k=1}^{n-j} W_k W_{k+j} = n\hat{w}_j(n) \quad \text{for } 1 \leqslant j \leqslant n-1.$$

Since $S_0(n)$ is a sum of i.i.d. centered random variables having variance 2 and arbitrary moments, the classical Hartman-Wintner law of the iterated logarithm implies

(106) $$\overline{\lim_{n\to\infty}} \frac{|S_0(n)|}{\sqrt{2n\log\log n}} = \sqrt{2}$$

which proves (102).

Fix $\varepsilon > 0$ and consider the stopping time

$$\tau_n = \inf\{r \geqslant 1 \mid S_0(r) > (1+\varepsilon)\sqrt{4n\log\log n}\,\}.$$

By (106) there is a **finite** random integer $N(\omega)$ such that P-a.s. one has

(107) $$\tau_n > n \qquad \text{for} \quad n \geqslant N(\omega).$$

Let $\mathcal{F}_n$ be the σ-algebra generated by $W_1 \ldots W_n$. For each fixed $j \geqslant 1$, $S_j(n)$ is an $\mathcal{F}_n$-martingale indexed by $n \geqslant j+1$. Moreover for $j \geqslant 1$, $n \geqslant j+1$, $u \in \mathbb{R}$ one has, since W is a gaussian white noise,

$$E\left[\exp\left(uW_{n-j}W_n - \frac{1}{2}u^2W_{n-j}^2\right) \Big| \mathcal{F}_{n-1}\right] \equiv 1$$

so that for all **fixed** $u \in \mathbb{R}$, $j \geqslant 1$, the process

$$M_j(n) = \exp\left[uS_j(n) - \frac{1}{2}u^2(S_0(n-j)+n-j)\right] \text{ where } n \geqslant j+1$$

is a positive $\mathcal{F}_n$-martingale with

$$E[M_j(n)] \equiv 1 \qquad \text{for } n \geqslant j+1,\ j \geqslant 1.$$

For any given $a > 0$, $j \geqslant 1$ define a stopping time $\gamma_j(a)$ by

$$\gamma_j(a) = \gamma_j = \inf\{r \geqslant j+1 \mid S_j(r) \geqslant a\}.$$

Let $c_n = (1 + \varepsilon)\sqrt{4n \log \log n}$. The event $\{\gamma_j(a) \leqslant n \wedge \tau_n\}$ obviously implies

$$\{S_j(n \wedge \gamma_j) = S_j(\gamma_j) = a \text{ and } S_0(n \wedge \gamma_j - j) \leqslant c_n\}$$

which in turn implies

$$F = \left\{M_j(n \wedge \gamma_j) \geqslant \exp\left[ua - \frac{1}{2}u^2(n - j + c_n)\right]\right\}.$$

Hence we have

$$P(\gamma_j \leqslant n \wedge \tau_n) \leqslant P(F)$$

$$\leqslant E[M_j(n \wedge \gamma_j)]\exp\left[-ua + \frac{1}{2}u^2(n-j+c_n)\right].$$

But γ_j being a stopping time, the process $t \to M_j(t \wedge \gamma_j)$ is still a martingale for $t \geqslant j + 1$ and $E[M_j(n \wedge \gamma_j)] = E[M_j[(j+1) \wedge \gamma_j]]$ $= E[M_j(j+1)] = 1$, whence the bound

$$P(\gamma_j \leqslant n \wedge \tau_n) \leqslant e^{-ua+ \frac{1}{2}u^2(n-j+c_n)}.$$

Take now $u = a/n - j + c_n$ to get

$$P(\gamma_j(a) \leqslant n \wedge \tau_n) \leqslant e^{-(a^2/2(n-j+c_n))} \quad \text{for } n \geqslant j+1,\ j\leqslant 1,\ a>0$$

Consider now for $a > 0$, $n \geqslant 1 + s$, the event

$$F(s,n,a) = \left\{\max_{1\leqslant j\leqslant s}\ \max_{1\leqslant r\leqslant n} S_j(r) \geqslant a\right\}$$

which obviously satisfies

$$[F(s,n,a) \cap \{\tau_n > n\}] \subset \bigcup_{1\leqslant j\leqslant s} \{\gamma_j(a) \leqslant n \wedge \tau_n\}$$

so that for all $a > 0$, $1 + s \leqslant n$,

$$P[F(s,n,a) \cap \{\tau_n > n\}] \leqslant se^{-(a^2/2(n-s+c_n))}.$$

Let n_k be an increasing sequence of integers with $\lim_{k\to\infty} n_k = +\infty$. Fix two constants $b,H > 0$ and let

$$s_k = b \log n_k \qquad a_k = H\sqrt{n_k \log \log n_k}.$$

We shall select below b, H and (n_k) such that

$$(108) \qquad \sum_{k\geqslant 1} s_{k+1}\exp\left[-\frac{a_k^2}{2(n_{k+1}-s_{k+1}+c(n_{k+1}))}\right] < +\infty.$$

Note that the use of a_k instead of a_{k+1} is crucial below! By Borel-Cantelli's lemma, P-a.s., there is a finite random integer $K(\omega)$ such that for **all** $k \geqslant K(\omega)$ the events

$$\{F(s_{k+1},n_{k+1},a_k) \cap (\tau_{n_{k+1}} > n_{k+1})\}$$

are **not** realized.

But, by increasing K if need be, we may of course impose $n_{K(\omega)} \geqslant N(\omega)$ where $N(\omega)$ has been introduced in (107). Then (107) shows, in view of the defining property of K, that P-a.s., for $k \geqslant K(\omega)$, the events $F(s_{k+1},n_{k+1},a_k)$ are *not* realized. This means that, P-a.s.

$$(109) \qquad \max_{1\leqslant r\leqslant n_{k+1}} \; \max_{1\leqslant j\leqslant b\log n_{k+1}} S_j(r) < H\sqrt{n_k \log\log n_k}$$

for $k \geqslant K(\omega)$.

Let now $K_1(\omega) = n_{K(\omega)}$. For any $n > K_1(\omega)$ we can find k such that $k \geqslant K(\omega)$ and $n_{k+1} \geqslant n > n_k$. In particular (109) holds, and *a fortiori* (taking $r = n$ in (109)) we get

$$\max_{1\leqslant j\leqslant b\log n} S_j(n) < H\sqrt{n_k \log\log n_k} < H\sqrt{n \log\log n}\,.$$

Consequently, as soon as we can exhibit b, H, n_k satisfying (108), we shall have proved the existence of a P-a.s. finite integer $K_1(\omega)$ such that P-a.s.,

$$(110) \qquad \max_{1\leqslant j\leqslant b\log n} S_j(n) < H\sqrt{n \log\log n} \quad \text{for all } n \geqslant K_1(\omega).$$

Now take $n_k = d^k$ with fixed $d > 1$. Then we have

$$s_k \sim s_{k+1} \sim k\, b \log d; \quad a_k^2 \sim H^2 d^k \log k$$

$$[c(n_{k+1}) + n_{k+1} - s_{k+1}] \sim n_{k+1} = d^{k+1}$$

so that the general term of the series (108) is equivalent to

$$\frac{b \log d}{k^{((H^2/2d)-1)}}\,.$$

Thus the series (108) converges as soon as $H^2/2d > 2$. Fixing any $H > 2$ we can always select $d > 1$ such that $H^2/2d > 2$. Then (108) converges and (110) holds.

Since $S_j(n) = n\hat{w}_j(n)$ for $j \geqslant 1$, we deduce from (110) that for any constant b, any $h > 1$ one has P-a.s.

$$\overline{\lim_{n\to\infty}} \max_{1\leqslant j\leqslant b \log n} \left[\frac{n}{2 \log \log n}\right]^{1/2} \hat{w}_j(n) \leqslant \frac{H}{\sqrt{2}} .$$

An identical argument yields the same result for the martingales $[-S_j(n)]$ which concludes the proof of Lemma 7.8.

It would be easy at this point to prove that actually (103) is also an almost sure equality but that is a consequence of the general Theorem 7.3. anyway.

7.9. *Individual Iterated Logarithm for White Noise Periodigrams*

Lemma. *Let W be a centered gaussian white noise, with variance σ^2 and global law P. Let $g \in C(\mathbb{T})$ be an arbitrary* **even,** *real valued function having exponentially decreasing Fourier coefficients. Then P-a.s.*

$$\overline{\lim_{n\to\infty}} \frac{n}{2 \log \log n}[I_n^W(g) - I^W(g)] = \sigma^2\left[\frac{1}{\pi} \int_{\mathbb{T}} |g(\lambda)|^2 d\lambda\right]^{1/2}$$

Proof. *With no loss of generality we may of course assume* $\sigma^2 = 1$. Let

$$M_n(g) = n[I_n^W(g) - I^W(g)] \text{ and } M_0(g) = 0.$$

From formulas (87), (104) and (105) we get

$$M_n(g) = \sum_{|j|\leqslant n-1} S_j(n)\hat{g}_j.$$

Since g is real-valued and even, we have $\hat{g}_j = \hat{g}_{-j} \in \mathbb{R}$ and we obtain readily

$$(111) \qquad M_n(g) - M_{n-1}(g) = (W_n^2 - 1)\hat{g}_0 + 2W_n \sum_{j=1}^{n-1} W_{n-j}\hat{g}_j$$

which forces $E[M_n(g) - M_{n-1}(g) | \mathcal{F}_{n-1}] = 0$. Thus $M_n(g)$ is a centered square integrable $\mathcal{F}_n$-martingale for fixed g. An elementary computation gives

$$E([M_n(g) - M_{n-1}(g)]^2 | \mathcal{F}_{n-1}) = 2\hat{g}_0^2 + 4\left[\sum_{j=1}^{n-1} W_{n-j}\hat{g}_j\right]^2 \qquad \text{for } n \geqslant 2.$$

Consider now the associated increasing process

(112)
$$U_n = \sum_{k=1}^{n} E([M_n(g) - M_{n-1}(g)]^2 | F_{n-1}) = 2n\hat{g}_0^2 + 4 \sum_{k=2}^{n} \left[\sum_{j=1}^{k-1} \hat{g}_j W_{k-j}\right]^2 .$$

Provided one can show that $\lim_{n\to\infty} U_n = +\infty$, P-a.s., and that one can check another technical assumption, then the law of the iterated logarithm for martingales [Stout [49], pp. 301-302] can be applied to $M_n(g)$ to yield

$$\overline{\lim_{n\to\infty}} \frac{M_n(g)}{\sqrt{2U_n \log\log U_n}} = 1 \qquad P\text{-a.s.}$$

We first study U_n for large n. Expanding the expression (112) of U_n, we get by direct reshuffling of sums

$$\frac{1}{4n} U_n = \frac{1}{2} \hat{g}_0^2 + \sum_{1\leq j,\ell\leq n} \hat{g}_j \hat{g}_\ell V_{j\ell}(n)$$

where

$$V_{j\ell}(n) = \frac{1}{n} \sum_{k=1+(j\vee\ell)} W_{k-j} W_{k-\ell}.$$

For $1 \leq j \leq \ell \leq n$ we may write

(113)
$$V_{j\ell}(n) = V_{\ell j}(n) = \frac{1}{n} \sum_{r=1}^{n-\ell} W_r W_{r+\ell-j} = \hat{w}_{\ell-j}(n) + \varepsilon_{j\ell}(n)$$

where the $\hat{w}_k(n)$ are the estimated covariances of W and $n\,\varepsilon_{j\ell}(n)$ is the sum of j random variables of the form $W_m W_{m'}$. Define now

(114)
$$R_n = \frac{1}{4n} U_n - \frac{1}{2}\hat{g}_0^2 - \sum_{1\leq j\leq n} \hat{g}_j^2 = \sum_{1\leq j\leq n} \hat{g}_j^2 (V_{jj}(n) - 1) + 2 \sum_{1\leq j<\ell\leq n} \hat{g}_j \hat{g}_\ell V_{j\ell}(n).$$

To study $\lim_{n\to\infty} R_n$ we proceed exactly as in Lemma 7.7 where the quantities $c_r c_s$, $F_{rs}(n)$, $G_{rs}(n)$ are replaced by $\hat{g}_j \hat{g}_\ell$, $V_{j\ell}(n)$, $\varepsilon_{j\ell}(n)$. Letting $b = 3\log(1/\beta)$ and assuming $g \in H(\beta,B)$, we first truncate at $b \log n$ exactly as in (95) and (96), after having noticed that $\|V_{j\ell}(n)\|_2 \leq 1$. This yields

(115)
$$R_n = \sum_{1\leq j<b\log n} \hat{g}_j^2 (V_{jj}(n) - 1) + 2 \sum_{1\leq j<\ell\leq b\log n} \hat{g}_j \hat{g}_\ell V_{j\ell}(n) + O_{a.s.}(1/n).$$

The cardinal of the set of products $W_m W_{m'}$ required to write down any single one of the $(n\ \varepsilon_{j\ell}(n))$ is $j < b \log n$, so that the set of all products $W_m W_{m'}$ required to write down all the $(n\ \varepsilon_{j\ell}(n))$ with $1 \leq j \leq \ell \leq b \log n$ is inferior to $(b \log n)^3$. We now get, exactly as in (92), (93) and (94) the bound

$$\sup_{1 \leq j \leq \ell \leq b \log n} |\varepsilon_{j\ell}(n)| = O_{a.s}\left[\frac{(\log n)^2}{n}\right].$$

By definition (113) of the $\varepsilon_{j\ell}(n)$, the expression (115) of R_n now becomes

$$R_n = (\hat{w}_0(n) - 1)\left(\sum_{1 \leq j < b \log n} \hat{g}_j^2\right)$$

$$+ 2 \sum_{1 \leq j < \ell < b \log n} \hat{g}_j \hat{g}_\ell \hat{w}_{\ell - j}(n) + O_{a.s.}\left[\frac{(\log n)^2}{n}\right].$$

By Lemma 7.8, we know that $|\hat{w}_0(n) - 1|$ and

$$\sup_{1 \leq s < b \log n} |\hat{w}_s(n)| \text{ are both } O_{a.s.}\left[\left(\frac{\log \log n}{n}\right)^{1/2}\right],$$

and hence

$$R_n = O_{a.s.}\left[\left(\frac{\log \log n}{n}\right)^{1/2}\right].$$

In particular $\lim_{n\to\infty} R_n = 0$, P-a.s. and in view of (114) we obtain P-a.s.

$$\lim_{n\to\infty} \frac{1}{2n} U_n = \hat{g}_0^2 + 2 \sum_{j=1}^{\infty} \hat{g}_j^2 = \frac{1}{2\pi} \int_{-\pi}^{\pi} |g(\lambda)|^2 d\lambda.$$

Thus P-a.s., $U_n \sim nc(g)$ where $c(g) = (1/\pi)\int_{\Pi} |g|^2 d\lambda$, and hence P-a.s.

$$\sqrt{2U_n \log \log U_n} \sim \sqrt{2n \log \log n}\,\sqrt{c(g)}\,.$$

The last point we need to check before applying the law of iterated logarithm for martingales is that the tail of $(M_n - M_{n-1})$ decreases fast enough (cf. [49], pp. 302-303).

But by the argument of (91) and (92) we have

$$P\left[\sup_{1 \leq t \leq n} |W_t| > 3\sqrt{\log n}\right] \leq \frac{1}{n^{7/2}}$$

so that Borel-Cantelli's lemma and (111) yield

$$M_n(g) - M_{n-1}(g) = O_{a.s.}(\log n)$$

which is much more stringent than needed in [49], pp. 302-303. Finally applying [49], pp. 302-303 we conclude that P-a.s.

$$\overline{\lim_{n\to\infty}} \frac{M_n(g)}{\sqrt{2n \log \log n}} = \left[\frac{1}{\pi} \int_{\mathbb{T}} |g|^2 d\lambda\right]^{1/2}$$

which achieves the proof of 7.9 since $\sigma^2 = 1$.

7.10. *Uniformity of the Iterated Logarithm*

Lemma. Let *W be a centered gaussian white noise. Fix an arbitrary compact* $H(\beta,B)$ *in* $C(\mathbb{T})$. *Consider the random maps* $g \to F_n^X(g)$ *where*

$$F_n^W(g) = \sqrt{\frac{n}{2 \log \log n}} \, [I_n^W(g) - I^W(g)].$$

Then P-a.s., *the sequence* F_n^W *is equicontinuous on* $H(\beta,B)$ *(endowed with the* $\| \, \|_\infty$ *topology).*

Proof. For $\varepsilon > 0$ set $K(\varepsilon) = \{g \in H(\beta,2B)$ such that $\|g\|_\infty \leq \varepsilon\}$. Consider the random variables

$$u(\varepsilon) = \sup_{n\geq 0} \sup_{g\in K(\varepsilon)} |F_n^W(g)|$$

which may a priori be infinite. For $h_1,h_2 \in H(\beta,B)$ and $\|h_1 - h_2\|_\infty \leq \varepsilon$, one clearly has $|F_n^W(h_1) - F_n^W(h_2)| \leq u(\varepsilon)$. To prove Lemma 7.10 we only need to show that $\lim_{\varepsilon\to 0} u(\varepsilon) = 0$, P-a.s.

We can of course assume var $W_n = 1$. With the notations of Section 7, we have, by (78)

$$(116) \qquad I_n^W(g) - I^W(g) = (\hat{w}_0(n) - 1)\hat{g}_0 + \sum_{1\leq|j|\leq n-1} \hat{w}_j(n)\hat{g}_j.$$

Let $\delta_j = 0$ for $j \neq 0$ and $\delta_0 = 1$. By Lemma 7.8 we know that the random variable

$$(117) \qquad w(n) = \sup_{0\leq|j|\leq b\log n} |\hat{w}_j(n) - \delta_j| \text{ is an } O_{a.s.}\left[\frac{\log \log n}{n}\right]^{1/2}.$$

On the other hand we have obviously

$$(118) \qquad \begin{aligned} &\sup_{g\in K(\varepsilon)} |(\hat{w}_0(n) - 1)\hat{g}_0 + \sum_{1\leq|j|<b \log n} \hat{w}_j(n)\hat{g}_j| \\ &\qquad \leq w(n)\Big[\sup_{g\in K(\varepsilon)} \sum_j |\hat{g}_j|\Big]. \end{aligned}$$

For $g \in K(\varepsilon)$ one has, for all $m \geqslant 0$

$$\sum_j |\hat{g}_j| \leqslant \sum_{|j|<m} |\hat{g}_j| + \sum_{|j|\geqslant m} |\hat{g}_j| \leqslant m\varepsilon + \frac{4B}{1-\beta}\beta^m.$$

Taking $m \sim \log \varepsilon / \log \beta$ we get immediately

$$\sup_{g\in K(\varepsilon)} \left[\sum_j |\hat{g}_j|\right] \leqslant c\sqrt{\varepsilon}$$

where c is a constant. From (117) and (118) we now deduce

$$\text{(119)} \qquad \sup_{g\in K(\varepsilon)} |(\hat{w}_0 - 1)\hat{g}_0 + \sum_{1\leqslant|j|\leqslant b\log n} \hat{w}_j \hat{g}_j| = \sqrt{\varepsilon}\, O_{\text{a.s.}}\left(\frac{\log\log n}{n}\right)^{1/2}$$

By Cauchy-Schwartz, we have the bound, valid for $1 \leqslant j \leqslant n-1$

$$|\hat{w}_{-j}(n)| = |\hat{w}_j(n)| = \left|\frac{1}{n}\sum_{k=1}^{n-j} W_k W_{k+j}\right|$$

$$\leqslant \frac{1}{n}\left(\sum_{k=1}^{n-j} W_k^2\right)^{1/2}\left(\sum_{k=j+1}^{n} W_k^2\right)^{1/2} \leqslant \frac{1}{n}\sum_{k=1}^{n} W_k^2 = \hat{w}_0(n).$$

which implies

$$\text{(120)} \qquad \begin{aligned} \left|\sum_{b\log n\leqslant|j|\leqslant n-1} \hat{w}_j(n)\hat{g}_j\right| &\leqslant \hat{w}_0(n)\left[\sum_{j\geqslant b\log n} |\hat{g}_j|\right] \\ &\leqslant \hat{w}_0(n)\,\frac{2B\beta^{b\log n}}{1-\beta}. \end{aligned}$$

But the law of large numbers (or (117) shows that $\hat{w}_0(n) = O_{\text{a.s.}}(1)$. Taking $b = 2\log(1/\beta)$ in (120), we get then

$$\text{(121)} \qquad \sup_{g\in K(\varepsilon)} \left|\sum_{b\log n\leqslant|j|\leqslant n-1} \hat{w}_j(n)\hat{g}_j\right| = O_{\text{a.s.}}(1/n^2).$$

Now (121), (119) and (116) imply immediately

$$\text{(122)} \qquad \begin{aligned} \sup_{g\in K(\varepsilon)} |F_n^W(g)| &= \left(\frac{n}{\log\log n}\right)^{1/2} \sup_{g\in K(\varepsilon)} |I_n^W(g) - I^W(g)| \\ &= \sqrt{\varepsilon}\, O_{\text{a.s.}}(1) + O_{\text{a.s.}}(1/n). \end{aligned}$$

Call $A_n(\varepsilon)$ the left-hand side of (122). By definition of $O_{\text{a.s.}}(\cdot)$, (122) provides a P-a.s. finite random variable C such that $A_n(\varepsilon) \leqslant C(\sqrt{\varepsilon} + 1/n)$ for all n.

Start now with an arbitrary deterministic number $\varepsilon_1 > 0$. We first compute and *fix* a *random* number ε, by $2C\sqrt{\varepsilon} = \varepsilon_1$, which implies

(123) $$\sup_{n \geq 1/\sqrt{\varepsilon}} A_n(\alpha) \leq 2C\sqrt{\varepsilon} = \varepsilon_1 \quad \text{for all } \alpha \leq \varepsilon.$$

For each fixed n, F_n^W is continuous on $C(\mathbb{T})$ and hence $\lim_{\alpha \to 0} A_n(\alpha) = 0$. This implies, since ε is fixed

(124) $$\lim_{\alpha \to 0} \left[\sup_{n \leq 1/\sqrt{\varepsilon}} A_n(\alpha) \right] = 0.$$

From (123) and (124) we now get a random α_1 such that $u(\alpha) = \sup_n A_n(\alpha) \leq \varepsilon_1$ as soon as $\alpha \leq \alpha_1$ which concludes the proof of Lemma 7.10.

7.11. Prooof of Theorem 7.3. Consider a deterministic sequence φ_n of continuous functions on an arbitrary compact K. Assume that the sequence is equicontinuous and that

(125) $$\overline{\lim_{n \to \infty}} |\varphi_n(g)| = \varphi(g) \quad \text{for all } g \in K,$$

where φ: $K \to \mathbb{R}$ is continuous.

Then the convergence (125) *must* obviously be *uniform* in $g \in K$.

This elementary deterministic result and Sections 7.9 and 7.10 prove Theorem 7.3 when $X = W$. In the general case, we apply 7.7. to obtain P-a.s., *uniformly* in $g \in H(\beta,B)$,

$$\overline{\lim_{n \to \infty}} F_n^X(g) = \overline{\lim_{n \to \infty}} F_n^W\left(\frac{2\pi}{\sigma^2} fg\right) = \sigma^2 \left[\frac{1}{\pi} \int_{\mathbb{T}} \frac{4\pi^2 f^2 g^2}{\sigma^4} d\lambda \right]^{1/2}$$

$$= \left[4\pi \int_{\mathbb{T}} f^2 g^2 d\lambda \right]^{1/2}$$

which proves Theorem 7.3.

Bibliographical Hints

The introduction of compensated log-likelihood is due to *Akaike* (1), (2), although the precise formulas he introduced actually lead to nonconsistent estimators. The essential result concerning compensated likelihood is due to *Hannan* (3). An interesting and compact essay on Hannan's results is due to *M. Bouaziz*, whom we have followed on several points. The theorems on Toeplitz forms are derived from *Hall-Heyde*.

The uniform iterated logarithm for the periodogram is inspired by *Hannan et altri* (cf. [28], 29]) and *An Hong Zhi et altri* [4]. An interesting point of view on asymptotic compensators can be found in *Rissanen*. The nonconsistency of Akaike's estimator is due to *Shibata*. We have made an essential use of the laws of iterated logarithm for martingales (*Stout* [49]).

Chapter XV
A FEW PROBLEMS NOT STUDIED HERE

In this textbook, we have deliberately left aside a few interesting problems, some of which have important practical and theoretical consequences. We briefly sketch a few of these questions and give the basic bibliographical references.

1. Tests of Fit for ARMA Models

Assume that we have identified an ARMA model and estimated its coefficients. We have thus obtained a model of the form

$$X_n + \hat{a}_1 W_{n-1} + \dots + \hat{a}_p X_{n-p} = W_n + \hat{b}_1 W_{n-1} + \dots + \hat{b}_q W_{n-q}$$

where p and q have also been estimated and hence are random variables which should have been denoted by $\hat{p}$, $\hat{q}$. We also have an estimate $\hat{\sigma}^2$ for the variance of the innovation W.

We now want to check the validity of the model by statistical methods. This is the purpose of the so-called **tests of fit.**

These tests generally derive from the following considerations. By backforecasting, we may obtain estimates of the "initial" values $\hat{X}_0$, $\hat{X}_{-1}$, ..., $\hat{X}_{-p+1}$ and $\hat{W}_0$, ..., $\hat{W}_{-q+1}$.

Starting with these values, the $\hat{W}_m$ can be obtained recursively in terms of the observations $X_1 ... X_n$. This method, described in Chapter 12, provides us with a sequence $\hat{W}_1 \dots \hat{W}_n$ of **estimated residuals**, whose values depend of course on the actual model $(\hat{a}_1 \dots \hat{a}_p\ \hat{b}_1 \dots \hat{b}_q,\ p,\ q)$ which we have selected initially. Most tests of fit rely on the idea of verifying if the sequence $\hat{W}_1 \dots \hat{W}_n$ does "look like" a sequence of observed values of the **white noise** W. In particular one may check the independence of the $\hat{W}_j$ (gaussian case) or the absence of significant correlations (general case).

Consider then the empirical estimators of the covariances of W, given by

$$\hat{w}_k = \frac{1}{n-p-k} \sum_{i=p+1}^{n-k} \hat{W}_i \hat{W}_{i+k} - \left[\frac{1}{n-p} \sum_{i=p+1}^{n} \hat{W}_i\right]^2 .$$

If the model is good, the $\hat{w}_k$ should be small.

The study of the law of $\hat{S}_k = (\hat{w}_1 \dots \hat{w}_k)$ is related to the study of $S_k = (w_1 \dots w_k)$ where

$$w_k = \frac{1}{n-k} \sum_{i=1}^{n-k} W_i W_{i+k} - \left[\frac{1}{n-k} \sum_{i=1}^{n} W_i\right]^2 .$$

We have seen in Chapter 10, Theorem 4.1 that $(1/\sigma^2)\sqrt{n}\ S_k$ is asymptotically normal, centered, and has an asymptotic covariance matrix equal to the identity matrix in dimension k. In particular $n\|S_k\|^2$ converges in law to χ^2_k.

The situation is more involved for

$$n\|\hat{S}_k\|^2 = n\left[\sum_{j=1}^{k} \hat{w}_j^2\right],$$

which converges in law to a χ^2_{k-p-q} *if p and q are assumed to be given and not estimated.* The proof is similar to the asymptotic study of χ^2 with estimated parameters (cf. [15], Vol. 2, Chapter 3).

Concerning this point, we refer the reader to *Box-Tiao.* These tests, which are known under the name of "tests of portmanteau" are also studied in *Box-Jenkins*, *Hannan*, and *Quenouille* who has introduced a test based on partial autocorrelations. The loss of degrees of freedom is due to the estimation of the $(p + q)$-parameters $\hat{a}_1 \dots \hat{a}_p$, $\hat{b}_1 \dots \hat{b}_q$.

Other tests have been proposed to check the "whiteness" (noncorrelation) of $\hat{W}_1 \dots \hat{W}_n$. For instance, if are sets

$$F_n(u) = \frac{1}{2\pi\sigma^2} \int_0^u I_n^W(s)\,ds,$$

where I_n^W is the periodogram of W, then

$$\sup_{s\in[-\pi,+\pi]} \sqrt{n}\,[F_n(s) - s]$$

has a limit in law (*Bartlett*). Replacing W_j by $\hat{W}_j$ and σ^2 by $\hat{\sigma}^2$, one can estimate I_n^W by $\hat{I}_n^W$ and $F_n(u)$ by the corresponding integral $\hat{F}_n(u)$, which yields another test of fit; $\hat{F}_n$ is called the estimated **cumulative periodogram**, and the same name is used for the test based on

$$\sup_{s\in[-\pi,\pi]} [\sqrt{n}(\hat{F}_n(s) - s)].$$

Essentially it may help to detect in the estimated spectrum the frequencies with high densities (points at which $F_n(s) - s$ is large). These frequencies indicate possible values for the periods of unduly discarded seasonal effects.

The a priori detection of the main periods for a seasonal trend in the model, a point of importance in seasonal ARIMA model building, has been widely studied. We refer to *Hannan* (1) and (2), *Walker*, *Bartlett*, on this point.

2. Nonlinearity

Linear stationary models and in particular ARMA models keep the *same* statistical structure *when the direction of time is reversed.* On the other hand some phenomena have trajectories which are clearly irreversible, for instance where the "ups" are steeper than the "downs". These phenomena cannot be modelized by ARMA models, except in some cases by applying nonlinear changes of coordinates to the observations.

Another important phenomenon is the following; when an ARMA model is used to prediction purposes, the k-steps prediction tends to zero when $k \to +\infty$. In other words, when one cancels the innovation considered at the exciting noise in the system, then the oscillations of the system dampen rather fast. However certain phenomena, placed in the same situation (lack of exciting noise), exhibit permanent random oscillations whose amplitudes do not dampen.

Some attempts have been made to study classes of nonlinear models which could exhibit these self-exciting oscillations, among other features. The theory at this point is not very satisfactory; the question of model identification is barely formalized and certainly not solved for most restricted classses of nonlinear models. We refer to *Priestley* (Vol. II) for a presentation of the main nonlinear models and a discussion of the gaussian hypothesis.

APPENDIX

1. Hilbert Spaces

1.1. Definition. A *complex Hilbert space* is a complex vector space E endowed with a map $(x,y) \to \langle x,y \rangle$ from $E \times E$ into $\mathbb{C}$, called *scalar product* and satisfying the following conditions.

(1) The map $x \to \langle x,y \rangle$ is, for all fixed $y \in E$, a linear map from E to $\mathbb{C}$.

(2) For all $x,y \in E$ one has $\langle y,x \rangle = \overline{\langle x, y \rangle}$.

(3) $\|x\|^2 = \langle x,x \rangle \geqslant 0$; $\|x\| = 0$ is equivalent to $x = 0$.

(4) Endowed with the distance $d(x,y) = \|x - y\|$ the space E is a complete metric space (i.e. every Cauchy sequence must converge in E).

1.2. Linear Forms, Weak Convergence

The set of continuous linear forms on E, that is of linear maps $z: E \to \mathbb{C}$ such that $z(x_n) \to 0$ whenever $\|x_n\| \to 0$, is a vector space E^*.

Each element $y \in E$ defines an element $y^* \in E^*$ such that $y^*(x) = \langle x,y \rangle$ for all $u \in E$.

Theorem. *To each continuous linear form z on E, one can associate a unique element y of E such that $y^* = z$.*

One may then identify E and its dual E^*.

Definition. One says that *a sequence* $y_n \in E$ *converges weakly to* $y \in E$ if for every $x \in E$, $<x,y_n>$ converges to $<x,y>$.

In particular, if H is a *closed* subspace of E, and if a sequence $y_n \in H$ converges weakly to $y \in E$, then one necessarily has $y \in H$.

2. Fourier Series; L^2 and ℓ^2-Spaces

Let ℓ^2 be the space of sequenes $(c_n)_{n\in\mathbb{Z}}$ with $c_n \in \mathbb{C}$, such that $\Sigma_n|c_n|^2 < \infty$. Endowed with the scalar product

$$<(c_n),(d_n)> = \sum_n c_n\bar{d}_n,$$

the space ℓ^2 is a complex Hilbert space.

Let L^2 be the space of complex valued functions f on $\mathbb{T} = [-\pi,\pi[$ such that $\int_{\mathbb{T}} |f(\lambda)|^2 d\lambda < \infty$. Endowed with the scalar product

$$<f,g> = \frac{1}{2\pi} \int_{\mathbb{T}} f(\lambda)\bar{g}(\lambda)d\lambda,$$

L^2 is a complex Hilbert space.

Let $f_n(\lambda) = e^{-in\lambda}$ for $n \in \mathbb{Z}$, $\lambda \in \mathbb{T}$. The closure of the linear envelope of the $(f_n)_{n\in\mathbb{Z}}$ is equal to L^2, i.e. every vector in L^2 is limit (in L^2) of at least one sequence of finite linear combinations of the f_n. One also says that the set $\{f_n\}_{n\in\mathbb{Z}}$ is total in L^2. Note that the (f_n) are pairwise orthogonal and that $\|f_n\| = 1$.

2.1. Proposition. *Let* $c = (c_n)_{n\in\mathbb{Z}}$ *be an arbitrary sequence of complex numbers. Define*

$$S_N(c) = \sum_{|n|\leq N} c_n f_n.$$

Then $S_N(c)$ *converges in* L^2 *if and only if* $c \in \ell^2$. *Let then*

$$S(c) = \lim_{N\to\infty} S_N(c).$$

Definition. If $g \in L^2$, *the nth Fourier coefficient* $\hat{g}_n$ *of* g *is*

$$\hat{g}_n = <g,f_n> = \frac{1}{2\pi} \int_{\mathbb{T}} g(\lambda)e^{+in\lambda}d\lambda.$$

We write $\hat{g} = (\hat{g}_n)_{n\in\mathbb{Z}}$.

2.2. Theorem. *The map $c \to S(c)$ is an isometry from ℓ^2 onto L^2 and satisfies*

$$g = S(\hat{g}) = \sum_n g_n f_n \quad \text{for all } g \in L^2.$$

Define the **Dirichlet kernel**

$$D_N(z) = \frac{1}{2\pi} \frac{\sin[(2N + 1)z/2]}{\sin(z/2)}$$

and the **Fejer kernel**

$$\Phi_N(z) = \frac{1}{2\pi N} \left(\frac{\sin Nz/2}{\sin z/2}\right)^2.$$

2.3. Theorem. *If g is of class C^2, then $S_N(\hat{g})$ converges to g as $N \to \infty$, uniformly and in L^2. Moreover, one has*

$$S_N(\hat{g}) = \int_{\mathbb{T}} D_N(z) g(x + z) dz$$

for all $g \in L^2$.

2.4. Theorem. *The Cesaro sums*

$$\sigma_N(g) = \frac{1}{N}[S_0(\hat{g}) + \dots + S_N(\hat{g})]$$

are given by

$$\sigma_N(g) = \int_{\mathbb{T}} \Phi_N(z) g(x + z) dz$$

and for any continuous function g on $\mathbb{T}$, $\sigma_N(g)$ converges to g uniformly and in L^2.

The following result estimates the speed at which the Fourier coefficients of a smooth function decrease to zero.

2.5. Theorem. *If g: $[-\pi,\pi] \to \mathbb{C}$ is of class p, then*

$$\overline{\lim_{n\to\infty}} |n|^p |\hat{g}_n| < +\infty.$$

3. Convergence in Distribution; Levy's Theorem

Let μ_k be a sequence of probability measures on $\mathbb{R}^k$; call $\varphi_n(v)$ the characteristic function of μ_n. Tight convergence has been introduced in Chapter 10, Section 1.2. If μ_n converges tightly to μ then $\lim_{n\to\infty}\varphi_n(v) = \varphi(v)$ where φ is the characteristic function of μ, and the convergence is **uniform** in $v \in \mathbb{R}^k$. The converse is Levy's theorem.

Theorem. *If for all* $v \in \mathbb{R}^k$, $\varphi_n(v)$ *converges to* $\psi(v)$, *and if the function* ψ *is continuous at* $v = 0$, *then* ψ *is the characteristic function of a probability* μ *on* $\mathbb{R}^k$ *and* μ_n *converges tightly to* μ.

Corollary. *If for all* v *in a neighborhood of* 0, $\varphi_n(v)$ *converges to* $\varphi(v)$, *and if* φ *is the characteristic function of a probability* μ, *then* μ_n *converges tightly to* μ.

Bibliographical Hints

We suggest, for instance, *Rudin* for most of the material covered in the appendix.

BIBLIOGRAPHY

[1] AKAIKE H. (1) - *A new look at statistical model identification, I.E.E.E. Trans. Auto. Cont. A.C.* 19(1974), 716-723.

[2] AKAIKE H. (2) - *Likelihood of a model and information criteria*, Journal of Econometrics 16(1981), 3-14.

[3] ANDERSON T. W. - Time Series Analysis, J. Wiley, 1980.

[4] AN HONG ZHI, CHEN ZHAO GUO, HANNAN E. J. - (1982), *Autocorrelation, autoregression, and autoregressive approximation*, Ann. Statistics 10, 926-936.

[5] ASTROM K. J., WITTENMACH D. J. - Computer Controlled Systems, Prentice Hall (1984).

[6] BARTLETT M. - Stochastic Processes, Cambridge University Press, 1955.

[7] BEGHIN B., GOURIEROUX Ch., MONTFORT A. - *Identification of ARMA model, the corner method*, in Time Series (O. Anderson ed.), North-Holland, 1979.

[8] BILLINGSLEY P. - Ergodic Theory and Information, J. Wiley, 1975.

[9] BILLINGSLEY P. - Probability and Measure, Wiley, 1979.

[10] BOX G., JENKINS G. - Time Series Analysis, Holden Day, 1976.

[11] BOX G., PIERCE D. - *Distribution of residual autocorrelation*, Journ. of the Am. Stat. Ass. 65(1970), 1509-1532.

[12] BOUAZIZ M. - *Identification de l'ordre et loi du logarithme itéré pour les formes de Toeplitz* - (à paraitre 1985).

[13] BREIMAN L. - Probability, Addison-Wesley, 1968.
[14] COURSOL J., DACUNHA-CASTELLE D. - *Remarques sur l'approximation de la vraisemblance des processus gaussiens stationnaires*, Teor. Veroy. Prim. 27(1982), 155-159.
[15] DACUNHA-CASTELLE D., DUFLO M. - Probability and Statistics, Vol. I and II, Springer-Verlag, 1986.
[16] DURBIN J. - *The fitting of time series models*, Rev. Int. Inst. Stat. 28(1960), 233-260.
[17] EYKHOFF P. - System Identification, Wiley, 1974.
[18] FRANKLIN G. F., POWELL J. D. - Digital Control, Addison-Wesley, 1981.
[19] GOURIEROUX C., MONTFORT A. - Series Temporelles, Economica, Paris 1981.
[20] GRANGER C. W. J., NEWBOLD P. - Forecasting Econometric Time Series, Academic Press, 1977.
[21] GRENANDER U., ROSENBLATT M. - Statistical Analysis of Stationary Time Series, Wiley, 1957.
[22] GRANTMACHER A. - Theorie des Matrices, Dunod, 1966.
[23] JOHNSON L. - Econometric Methods, McGraw Hill, 1979.
[24] HALL P., HEYDE C. C. - Martingale Limit Theory and Its Application, Academic Press, 1981.
[25] HANNAN E. J. (1) - Multiple Time Series, Wiley, 1970.
[26] HANNAN E. J. (2) - *Nonlinear time series regression*, J. App. Prob. 8(1971), 767-780.
[27] HANNAN E. J. (3) - *The estimation of the order of an ARMA process*, Ann. Stat. Vol. 8, No. 5 (1980), 1071-1081.
[28] HANNAN E. J., KAVALIERIS L. - *Multivariate linear time series models*, Adv. Appl. Prob. 16(1984), 492-501.
[29] HANNAN E. J., RISSANEN J. - *Recursive estimation of ARMA order*, Biometrika 69(1982), 81-94.
[30] HOEL P., PORT S., STONE Ch. - Introduction to Stochastic Processes, Houghton-Mifflin, 1972.
[31] HOFFMAN V. - Banach Spaces of Analytic Function, Prentice Hall, 1962.
[32] IBRAGUIMOV I. - *On the estimation of the spectral function of a stationnary gaussian process*, Theory of Prob. and its App., Vol. III(1963), 366-400.
[33] IBRAGUIMOV I. A., ROZANOV Y. - Processus Aleatoires Gaussiens, Mir, 1974.
[34] JENKINS G. M., WATTS D. G. - Spectral Analysis and Applications, Holden Day, 1969.

[35] KAILATH T. - Linear Systems, Prentice Hall, 1980.
[36] KOOPMANS L. H. - The Spectral Analysis of Time Series, Academic Press,
[37] LOEVE M. - Probability Theory, 2nd Edition, Van Nostrand, 1972.
[38] MONTGOMERY D., JOHNSON L. - Forecasting and Time Series Analysis, McGraw Hill, 1976.
[39] NEVEU J. (1) - Processus Gaussiens, Presses de Universite de Montréal, 1968.
[40] NEVEU J. (2) - Bases Mathématiques du Calcul des Probabilités, Masson, 1970.
[41] OHLSEN R. A. - *Asymptotic properties of the periodogram of discrete stationary process*, J. Appl. Prob. 4(1967), 508-528.
[42] QUENOUILLE M. H. - *Approximate tests of correlation in time series*, Jour. Roy. Stat. Soc. B 11(1949), 68-126.
[43] PRIESTLEY M. B. - Spectral Analysis and Time Series, 2 Vols. Academic Press, 1981.
[44] RISSANEN J. - *Modelling by shortest data description*, Automatica 14(1978), 465-471.
[45] RISSANEN J. - *Universal prior for parameters and estimators by minimum description length*, Ann. Stat. 11(1981), 416-431.
[46] ROZANOV Y. - Processus Aléatoires, Editions de Moscou, 1975.
[47] RUDIN W. - Analyse Réelle et Complexe, Masson, 1975.
[48] SHIBATA R. - *Selection of the order of an autoregressive model by Akaike's criterion*, Biometrika 63(1978), 117-126.
[49] STOUT W. F. - Almost Sure Convergence, Academic Press, 1974.
[50] VENTSEL H. - Théorie des Probabilités, Editions de Moscou, 1977.
[51] WALKER G. - *On periodicity in series of related terms*, Proc. Roy Soc. Lond. A 131(1931), 518-532.
[52] WIDOM H. *Asymptotic inversion of convolution operator*, Annales IHES (1979), 191-240.
[53] WILDE D., BEIGHTER C. - Foundations of Optimization, Prentice Hall, 1967.
[54] YAGLOM A. M., YAGLOM I. M. - Probabilité et Information, Dunod, 1969.

INDEX